The Regulation of Water and Waste Services

The Regulation of Water and Waste Services

An Integrated Approach (RITA-ERSAR)

Edited by
Jaime Fernando de Melo Baptista

Entidade Reguladora dos Serviços de Águas e Resíduos
The Water and Waste Services Regulation Authority

Published by   **IWA Publishing**
Republic - Export Building
1 Clove Crescent
London E14 2BA, UK
Telephone: +44 (0)20 7654 5500
Fax: +44 (0)20 7654 5555
Email: publications@iwap.co.uk
Web: www.iwaponline.com

First published 2014
© 2014 IWA Publishing

British Library Cataloguing in Publication Data
A CIP catalogue record for this book is available from the British Library

ISBN 9781780406527 (Hardback)
ISBN 9781789065220 (Paperback)
ISBN 9781780406534 (eBook)

Contents

Chapter 5
Integrated regulatory approach ... 61

Chapter 6
Regulatory contribution to the organisation of the sectors ... 69

Chapter 7
Regulatory contribution to the legislation of the sectors ... 73

Chapter 8

Chapter 9

Chapter 10

Chapter 11

Chapter 12
Quality of service regulation **139**

Chapter 13
Drinking water quality regulation **177**

Chapter 14
User interface regulation **193**

Foreword from an international perspective

Having watched the evolution of regulatory theory and practice over the past thirty years in the context of electricity, solid water, water supply and waste water management in countries around the world, it is indeed a great honour to be able to write this International Forward to Jaime Melo Baptista's comprehensive new book 'The Regulation of Water and Waste Services – An Integrated Approach (RITA/ERSAR)'.

My mission in this forward is to not to replicate the outstanding National Perspective forward presented by the Mr. Amílcar Theias, the Former Minister of the Environment and Spatial Planning for Portugal, nor the excellent introduction that kicks off the book.

Instead, I would like to expand on why this story about the creation of the regulatory model RITA/ERSAR and unrelenting commitment of Jaime Melo Baptista, his staff, some of the major entities like Águas de Portugal and Empresa Portuguesa das Águas Livres and finally super-dedicated key political figures were fundamental to the creation of what I have publicly described as the 'Portuguese water miracle'. Let me explain.

Less than two decades ago, Portugal could be characterized almost as a developing country in terms of the basic supply of water, waste water and solid waste services. Only 50% of the publicly supplied water was regarded as safe to drink. Today this number is in excess of 98%. Less than two decades ago, only about 60% of waste water was collected and less than 30% treated. Today these numbers are both on their way to 85-90%. Finally, in less than two decades, there has been a complete transformation of the solid waste system from mostly uncontrolled dumps to a modern, sanitary and efficient solid waste system throughout the country.

Through this period and largely behind the scenes, there was the unfolding of a comprehensively conceived but evolutionarily implemented restructuring of the whole water and waste sectors involving national, regional, municipal levels and both public and private entities. Today, Portugal can be extremely proud of this level of institutional change although, like in even the best countries, it is a work in progress. And in contrast to far-wealthier countries like the United States of America and Germany, the change that I am describing was accomplished very cost effectively and had to survive a huge downturn in the national economy beginning in 2008/9.

This takes us to the best part of the story and the object of this book, the role of the national regulator in the realization of the 'Portuguese water miracle'. From the beginning, it became apparent to an outsider like myself that part of this water and waste miracle is the product of the very unique role played by ERSAR, compared with many other regulators around the world.

Under Jaime Melo Baptista's leadership, with critical support from a number of leaders in Portugal, ERSAR forged a cooperative rather than an adversarial relationship with those they were trying to change and in the process brought in the national associations. Some would say this is impossible because regulators should never be too close to those they regulate. But I would argue that the huge transformation of the Portuguese water and waste sectors would have been impossible without the carefully constructed balance within ERSAR between supporting this transformation and setting up a rigorous standards framework and associated enforcement mechanisms. And also impossible without the continued national leadership role played by ERSAR in working with all parts of the Portuguese water and waste sectors.

It has been a delight to witness, and at times contribute, to this national transformation within Portugal and I am so pleased to see that this book has been developed to document (in outcomes and methods) what is truly good news story in services that are so basic to everyday health and happiness of the society.

Paul Reiter
Former Executive Director at International Water Association
Lisbon, September 2014

Foreword from a national perspective

It was with immense pleasure that I accepted the invitation from Jaime Melo Baptista, President of the Water and Waste Services Regulation Authority in Portugal, to write the foreword to this book which is now being published with the title 'The Regulation of Water and Waste Services – An Integrated Approach (RITA-ERSAR)'. This publication arises at a particularly appropriate moment as it coincides with the World Water Congress, taking place in Lisbon, which is organised by the International Water Association (IWA) and whose president is the author of this book. In this work, Jaime Melo Baptista presents a regulation model inspired on what was designed and implemented in the Portuguese context during the transition from the twentieth to the twenty-first century. This was indeed a remarkable period for the development of water and waste services in Portugal. Whilst not seeking to be a treatise on regulation, the book extensively covers all the areas that should be developed by a regulator in the water and waste sectors. This is the result of the author's in-depth theoretical and practical knowledge and his extremely rich experience in this field throughout the world. The author, a civil engineer by training with a distinct career in the field of hydraulics and sanitary engineering, and who was distinguished with the 2012 IWA Award for Outstanding Contribution to Water Management and Science, admits, in the very first lines, that he has written this book with the primary aim of recording and transmitting the experience he has acquired as the chairperson of the Portuguese Water and Waste Services Regulation Authority (ERSAR), a fascinating period of his professional life.

Jaime Melo Baptista personifies the ideal of a great servant of the State, one who ensures the continuity of public service despite changes in policies, governments and ministers succeeding one another. Nevertheless, a great servant of the State means more than mere continuity and less than just tagging along. It implies making the general interest prevail, stimulating individuals and structures with great rigour and a spirit of openness, managing to disentangle difficult situations, reconcile and utilise stakeholders with original thinking and major independence of spirit. These are the attributes that I have found in Jaime Melo Baptista, attributes with which I completely feel identified and which are the sign of an independent regulator. As the Portuguese Minister of Environment Jorge Moreira da Silva recently stated, Jaime Melo Baptista has never needed a legal framework to be independent and exercise his influence.

However, the existence of a legal framework ensuring this independence is clearly an essential asset for any regulator. As such, the present moment is particularly gratifying for the author as it coincides, with an interval of just a few months, with the establishing in Portugal, by Law of the Parliament, of an independent administrative authority, empowered with real powers to regulate drinking water supply, waste water management and solid waste management services, in models very close to those which he has always argued for. This has been the culmination of around twelve years of very intensive work encompassing an unshakeable determination that sought to demonstrate to the different stakeholders within the sectors (government, municipalities, utilities, economic agents, environmentalists, and users of the services), through intense pedagogic activity, the benefits which would stem for all with the creation of independent and effective regulation. Making use of an organisation limited in resources and without the formal powers which would enable him to act as a true regulator, he has sought to overcome these restrictions through his moral independence allied with an extraordinary technical competence, with a working capacity that could be described as inexhaustible, and with an ability to listen and to argue without becoming authoritarian. These characteristics have always come alongside with the enthusiasm and energy which he has always managed to transmit to his interlocutors and collaborators within a spirit of constant initiative. In this regard, the words of Leonardo da Vinci, who was also a hydraulic engineer, can be recalled and applied to the character of Jaime Melo Baptista: '*I love those who can smile in trouble, who can gather strength from distress and grow brave by reflection*'. These qualities have enabled him, within a traditionally interventionist political climate and in a country where an economic market culture is not strongly rooted, to dramatically alter the perspective of the government and the sectors in general in relation to the necessary role of the organisation he oversees since 2003. The action of the regulator, during the period which has now terminated, was not carried out in an arbitrary manner, according to convenience or determined by circumstances, but rather consistently and systematically based on a conceptual matrix which has been developed and perfected, the RITA-ERSAR regulatory model, which the reader will find as the central theme of this book now being published, and the subject of detailed explanation herein, to quote the author, '*an integrated regulatory approach for the water and waste services. (. . .) This model is based on two major levels of intervention: structural regulation of the sectors and behavioural regulation of the utilities. The components of the structural regulation of the sectors consist of contributions towards the organisation, legislation, information and capacity building of the sectors. The components of the behavioural regulation of the utilities consist of legal and contractual regulation, economic regulation, quality of service regulation, drinking water quality regulation and user interface regulation. All these components must be perfectly linked to each other, so as to form a coherent and integrated model, the effectiveness of which depends from the synergies obtained between these components*'.

Of particular interest is the final chapter of the book, which presents a set of general recommendations addressed to the political powers, the utilities, regulatory authorities, users and society in general. In particular, the author recommends that '*Political power, as a key stakeholder in these sectors, ensures the implementation of appropriate public policies for the provision of the drinking water supply, waste water management and solid waste management services provided to the population, in particular within the framework of the Millennium Development Goals, which set targets for the water services in terms of population coverage, and the recent United Nations resolution, which declared water and sanitation services to be human rights. These policies should be based on an integrated approach involving various aspects, particularly through the definition of a tariff policy for services that promotes a gradual recovery trend for costs, compatible with the economic capacity of the population and which protects the most economically disadvantaged*'.

Jaime Melo Baptista concludes by expressing his profound conviction that '*that if the political powers implement a suitable public policy that includes an integrated regulatory approach, if the utilities and regulatory authorities perform their activity appropriately and if users play their part in proactive citizenship, then the essential conditions have been met to promote the provision of public water and waste services with universal access, continuity and quality and efficiency and price equity, thus constituting an important factor for social and economic development'.*

And this conviction, as shown by the current regulator, is halfway towards the goal of successful regulation. Quoting the English thinker and liberal economist John Stuart Mill, *'One person with a belief is equal to ninety-nine who have only interests'.*

In providing a (necessarily incomplete) summary of what have been the significant aspects of regulatory intervention in recent years in Portugal, and based on the approach developed in this book, Jaime Melo Baptista highlights the following:

- Invaluable support and ongoing dialogue with legislators and political stakeholders in drawing up a legislative and institutional framework for the sectors. A dialogue which is fundamental to cope with different idiosyncrasies, ever changing interests, and doctrinal preconceptions, strains that besides creating instability, often obliterate the policy maker's room for manoeuvre.
- The recent approval of the first tariff regulation, at present limited to the solid waste management sector, has thus established the provisions applicable to the definition, calculation, revision and publication of tariffs and the respective obligations concerning information provision. This was a major step in terms of greater transparency in establishing prices for services and which will lead, I am sure, to increased rationality in utilities management and to fairer conditions for user access to the service.
- The insistent contribution towards attaining results in terms of drinking water quality, unimaginable a dozen years ago in Portugal, has allowed the whole country to have safe drinking water nowadays.
- The crucial development of the process of understanding the utilities themselves, particularly their performance levels, seeking to stimulate a climate of healthy emulation and the search for improvements in efficiency, including through the awarding of awards for service quality by suitably qualified juries.
- Information creation and dissemination on a wide level regarding the sectors, thus allowing for updated analyses and ongoing interactivity with the main stakeholders.

I am certain that this work will become an invaluable reference work, not only for academics within the sectors, but above all as a guide to all those who hold responsibilities in the most varied bodies with interests in this area, particularly public or private service operators, and responsible politicians and administrators. This contribution seems particularly opportune in Portugal, where we are witnessing a turning of the page regarding public policies targeted at these sectors. After a particularly intense initial stage in terms of establishing infrastructure, a new stage unfolds, more focused on the search for an efficient utilisation of resources and growing awareness in terms of social and environmental responsibility, making the activity of the regulator even more necessary.

Finally, two brief notes. Firstly, one of congratulating Jaime Melo Baptista for the enormous quality of the work he presents in this publication, a clear demonstration of his commitment towards his profession and the result of many hours taken from his already limited spare time. Secondly, that of recording the generous service provided to society which this work constitutes and which will certainly be unanimously recognised as such.

Amílcar Theias
*Former Minister of the Environment and Spatial Planning of Portugal and
Chairperson of the Advisory Council of ERSAR
Lisbon, September 2014*

Author's note

I have written this book with the main aim of recoding and conveying the experience acquired during a fascinating period of my professional life as chairperson of the Water and Waste Services Regulation Authority (ERSAR) in Portugal, which I have carried out since 2003, now totalling nearly twelve years. I do this in the expectation that this experience can be helpful to the outstanding team with whom I have worked, to the future generations of my country and also to other countries whose development may benefit from this experience. I am doing this as a testimony to an exceptional period for water and waste services and for regulation in Portugal, in which I have had the opportunity and privilege to actively participate.

Having come from the world of scientific research, this period gave me the opportunity to put into practice a complete set of ideas that theoretical and conceptual reflection enabled me to develop, and demonstrate that research results can and should be transformed into effective development for the benefit of society.

My regulatory functions have taken place in a no less captivating period for the implementation of new public policies in Portugal for these essential public services, which began in 1993 and which has enabled the country to evolve dramatically over the past two decades. I have also been both a viewer in some periods and someone closely involved in implementing this policy in other periods, in which regulation has had a crucial role through being one of its most important drivers.

As far as the drinking water supply service is concerned, two decades ago around 81% of the population was covered with public services, and this percentage has continually and significantly increased to now stand at 95%, which is the national target. In relation to the quality of drinking water, in 1993 only about 50% could be considered safe, this indicator is now above 98%, very close to the target of 99%. This demonstrates excellent water quality which has resulted in an obvious positive impact in public health.

Two decades ago the waste water management only covered with public services 60% of the population with regard to collection and 28% with regard to treatment, and these proportions since then have been increasing continuously and significantly, currently reaching 81% and 79%, respectively. These figures are still far from the target of 90%, but they constitute a huge advance and consequently an extraordinary environmental improvement. This improvement can be seen, for example, in the quality of surface and bathing waters and the growing number of blue flags on our beaches.

Two decades ago, solid waste was largely carried out in an unsuitable manner, with the proliferation of uncontrolled dumps. This resulted in serious problems for the environment and public health. Currently there is almost full population coverage of the solid waste management service, all of which is forwarded to a suitable destination. Recycling and disposal infrastructure have been set up and separate collection systems for a variety of materials created, allowing the total closure of these dumps.

Alongside this development, the regulation of water and waste services in Portugal had historically been directly carried out by the State, in what may be called implicit regulation, performing one of its traditional functions. Only in 1993, when formulating new policies for these sectors, was there an understanding and an evolution towards more intense and sophisticated regulation, in what may be called explicit regulation, carried out by a specialised regulator and integrated within these new public policies.

The first step in the explicit regulation of these services was taken in 1995 with the creation of the Concessions Monitoring Committee. This committee had powers delegated by the State, being responsible for issuing opinions on the investment plans of new operating companies for multi-municipal systems and the tariffs they propose.

In the same year the National Observatory for Multi-municipal and Municipal Systems was set up which, however, was never operationalised. It was later wound up, in the difficult way of the regulation.

The Regulatory Office for Water and Waste (IRAR) was created in 1997 with the function of regulating those services on mainland Portugal. However, it only began to operate in 2000. The scope of its regulation only covered State and municipal owned systems operated under concession, except for the role of competent authority for controlling the quality of drinking water, which covered all water supply utilities, regardless of their management model.

In 2006, IRAR gave way to the Water and Waste Services Regulation Authority (ERSAR). The organic law for ERSAR was only passed in 2009, which extended regulation to all water and waste service utilities, regardless of management model. More recently, ERSAR changed from a public institute to an independent administrative authority in 2014 and increased its regulatory powers. Currently it regulates approximately five hundred State and municipally owned utilities, with direct, delegated and concessionaire management models.

I became chairperson of the regulator in 2003 when it was still the IRAR. During this time I have worked with five governments and eight ministers with responsibility for the environment, from three different political parties. The primary task was to design a regulatory model for the sectors that would enable an integrated approach and begin the difficult process of its gradual implementation and consolidation, in compliance with good regulatory practices. It has been possible to do so with great stability, despite the different political cycles of the country.

It is inspired in this regulatory experience that I describe here an integrated regulation model RITA-ERSAR for drinking water supply, waste water management and solid waste management services, which I believe will be applied for many years in Portugal and which may be applicable in other countries.

Jaime Melo Baptista
Chairperson of ERSAR, the Water and Waste Services
Regulation Authority in Portugal
Lisbon, September 2014

Acknowledgements

I would like to thank my family who have always given me all their support. To my parents Jaime and Maria José, who taught me to live with dignity based on sound ethical and moral values that I have kept throughout life and have sought to pass on to my daughters and my grandchildren. To my loved wife Dora, with whom I have shared nearly forty years of life and who always gave me support in the difficult balancing act between private life and public activity. To my beloved daughters Paula and Rita, who gave meaning to my life and shared fantastic years with us that nothing can erase, allowing me to strengthen feelings that words cannot explain. You, Rita, who inspired me in a very special way in the writing of this book. To my dear grandchildren Leonor and Miguel, fruit of the wedding of Paula with my son-in-law Filipe, who are my inspirations for the future and the passion of the present. And who have in their own way cooperated in this book, often playing around me as I tried to write a few more pages. To my brother António, to his wife Izildinha and my nephews Gustavo and Rafael, exemplary in their unconditional friendship and continued support through good and bad times. To my parents-in-law Antero and Amália, that I cannot forget.

I would like to thank my colleagues at ERSAR for all their years of hard and high quality work and the great dedication shown in such a continuously high-pressured and demanding environment. Only by this means has such consistent growth of regulatory intervention for water and waste services in Portugal been possible. This book is also their book. I wish to take this opportunity to encourage them to promote the on-going improvement and innovation of regulation and strengthen their team spirit in this period of great challenges and huge restrictions, for the benefit of Portuguese society. They must know they have the strength to believe that change is possible, and that it is necessary to improve.

I would like to mention here, in particular, the directors and coordinators of ERSAR departments, with whom I have naturally had a closer relationship, specifically Alexandra Cunha, Alexandre Milheiras Costa, Cecília Alexandre, Conceição Ribeiro, Cristina Aleixo, David Alves, Filomena Lobo, Isabel Andrade, João Almeida, Luís Engrossa, Luís Simas and Paula Freixial. I would also like to express my appreciation to some of them for reading, making comments and supporting the edition of this book at its review stage, particularly Álvaro Carvalho, David Alves, Filomena Lobo, Gisela Robalo, Isabel Andrade, João Rosa, Luis Engrossa, Luís Simas and Paula Freixial.

I also make a very special mention to my dedicated and competent secretary Cristina Pereira, undoubtedly my model of a perfect secretary.

I acknowledge here the work of my fellow members of the board in these last twelve years, Dulce Álvaro Pássaro, Rui Ferreira dos Santos, João Simão Pires, Fernanda Maçãs and Carlos Lopes Pereira, with whom I have shared the main regulatory decisions.

I also wish to acknowledge the chairperson of the ERSAR Advisory Council, Amílcar Theias, who has managed this body of major importance for the sectors so well and with such skill and dedication, promoting intense debates about the most important aspects of the sectors and regulatory instruments. I also wish to express my thanks for the friendship that he has always expressed since he was Minister of the Environment and Spatial Planning, in the 15th Constitutional Government, where we had a strong and fruitful institutional relationship.

I also wish to acknowledge the Sole Auditor, Maria do Rosário Líbano Monteiro, who has always carried out her duties in a diligent and friendly manner.

I want to express my appreciation and respect to Maria Teresa Pinto Basto Gouveia, ex-Secretary of State of Environment (1991–1993) and ex-Minister of Environment (1993–1995) of Portugal, who promoted with a great vision the implementation of new public policies for those water and waste services.

I would like to thank Pedro Cunha Serra, the first chairperson of the regulatory authority between 1999 and 2001, always committed and competent, and the author of a first strategic reflection on how to carry out the regulation of these services in Portugal. To Jorge Vasconcelos, distinguish chairman of the Portuguese Energy Services Regulatory Authority (ERSE) between 1997 and 2006, who gave me so important knowledge and advices.

I would also like to thank the many stakeholders in the sectors and the exemplary relationship that has always existed, particularly with municipalities, associations of municipalities, municipal and intermunicipal owned companies, State owned companies and private companies. I also praise the excellent relationship with user associations, economic and business associations, environmental protection associations and the media. Furthermore, good relationships were achieved with the public administration related to water and waste services, research and development centres, analytical laboratories, consulting and design companies, audit companies, quality management companies and manufacturers and suppliers of materials, equipment and products.

I would also like to thank José Eduardo Martins, Secretary of State for the Environment of the 15th Constitutional Government, for his confidence when on that day in late 2002 he telephoned to sound me out about my willingness to become the chairperson of the regulator. It was a time when I wanted to slow down the pace of my professional life a little, but where, once again, I could not resist an exciting new challenge.

I also thank my colleagues and friends of the national technical community, and there are so many, who have accompanied me throughout these almost four decades of activity, and who form my professional family, especially in the National Laboratory for Civil Engineering. I highlight the exceptional relationship of cooperation and friendship that I have always maintained with Helena Alegre, a scientific researcher with wide national and international recognition, whose research work has always inspired me. I would also highlight Fernando Abecasis, my head of the Hydraulics Department at LNEC, Armando Lencastre and Pedro Celestino da Costa, all those great leaders and friends, with whom I learnt so much.

I would like to thank my colleagues and friends at Engidro, Ambitec and Engiform, successful business initiatives of a group of young engineers, with whom I have shared more than a decade of interesting activity.

To my colleagues and friends in the international community, especially the European, American and African continents, with whom I have shared so many good times during international events and projects. You are also part of my professional family. Let me refer two very special friends with an outstanding and holistic view about the water problems around the world and a very special, constructive and friendly way to cooperate with others at international level. Paul Reiter, former Deputy Director for Strategic Policy at Seattle Public Utilities and former Executive Director at International Water Association between 2002 and 2012. And Mike Rouse, former Chief Executive and Managing Director of the Water Research Centre (WRc), former Head of Drinking Water Inspectorate in United Kingdom, past president of the International Water Association and Distinguished Research Associate at the University of Oxford. People like them help us to have a better world.

To the International Water Association, whose intense activities I have followed for over thirty years and which has always been the international reference platform for me in terms of contacts with the scientific and technical community.

Jaime Melo Baptista
Chairperson of ERSAR, the Water and Waste Services
Regulation Authority in Portugal
Lisbon, September 2014

About the author

The author, Jaime Fernando Melo Baptista, was born in 1953 in Luanda, at the time the Portuguese overseas province of Angola. He lived there for nearly two decades.

He completed a degree in Civil Engineering in 1975 at the Faculty of Engineering of the University of Porto, in Portugal, with a final average of 16 points (out of 20). He received the Engineer António de Almeida Foundation Award for achieving the highest mark that year, jointly with a few other students.

After a brief period working for Loures Municipality, on the infrastructures department, he began working at the National Laboratory of Civil Engineering (LNEC) in 1976. He then did the Specialization Course in Sanitary Engineering in 1976/77 at the New University of Lisbon.

He began his research career at LNEC as a Trainee Specialist (1976). He gradually moved up the ranks, from Research Assistant (1980), Assistant Researcher (1983) and Main Researcher (1988), to reach the highest level, Research-Coordinator, in 1994.

In LNEC he managed the Sanitary Hydraulics Division between 1984 and 1989, which consisted of a team of about 15. This division conducted research at national and international level in the fields of drinking water supply, waste water and storm water services.

Later, between 1990 and 2000, he headed the Hydraulics Department of LNEC, with a team of about 120, carrying out research and technological development in the field of water. It was the largest research unit of this area in the country, with expertise in Civil, Chemical and Environmental Engineering, Physical and Chemical Sciences, Mathematics, Geology and Biology. The main areas of activity were water resources such as rivers, groundwater, estuaries, the coastline and sea, river hydraulic structures, such as dams and weirs, marine hydraulic structures such as harbours, port structures, submarine structures, coastal defence structures and navigation channels, and basic sanitation systems, such as facilities for drinking water supply, waste water, storm water and solid waste management, and also the environment. Scientific work with these various activities utilised analytical studies, mathematical modelling, physical modelling, field observations and laboratory experimentation.

He carried out numerous research projects, particularly in the areas of mathematical modelling, rehabilitation, performance assessment and improvement in the quality of water supply systems. He carried out legislation and standardisation activities, scientific and technical training, and also the preparation of technical publications, tests, certifications and advisory reports. The theses he presented for public discussion in 1983, focusing on the economic design of water distribution systems, and in 1994 on a methodological approach to the rehabilitation of water distribution systems, are of note. Another highlight is the coordination of the project in 1994 on tools to support a policy of sustainable development of basic sanitation in Portugal, requested by the then Directorate General of the Environment and financed by the European Commission Cohesion Fund, which encompassed 17 sectoral studies and a team of close to 50 specialists. This study involved describing the situation of basic sanitation in the country, the institutional framework, financial and tariff instruments, the organisation of services, market conditions, delegated and concession management, legislation, technical standardisation, construction and operational costs, materials and facilities, analytical laboratories, performance indicators, statistical information, human resource training and research and development.

He is author or co-author of more than five hundred publications and technical and scientific papers, including theses, technical books, technical manuals, papers presented at national and international conferences, articles in national and international technical magazines, journal articles, technical standards, reports and advisory reports.

He has been a speaker at more than two hundred fifty national and international conferences. He has organised around eighty technical and scientific conferences, including many international ones. He has been the president of the World Water Congress organised by the International Water Association (IWA) in Lisbon in 2014, with about 5000 participants from one hundred countries. He organized the First International Forum for Water Services Regulators, as part of that World Congress, which brought together many regulators from around the world and contributed to the dissemination and harmonisation of best regulatory practices.

He has carried out various teaching activities and has been the academic supervisor for more than twenty doctoral and masters students, and been on more than thirty university juries and on seven scientific review panels.

As a representative of LNEC, he was part of the European Group of Hydraulics Laboratories and the Group of Research Institutes in the field of water between 1990 and 2000.

He was chairperson of the Portuguese Committee for basic sanitation standardisation from 1990-1994, participating and monitoring in particular the drawing up of European legislation.

He has been a member of the Engineering Academy since 1997 and was a member of its board of directors between 1998 and 2000, becoming a member once more in 2010. He has been a member of the Portuguese Engineering Council, as a Senior Member, since 1997, and as an expert in Sanitary and Hydraulic Engineering and Water Resources since 1998, and was appointed an Advisory Member in 2003. He was a founding member of the Portuguese Association of Water Resources and was Vice Chairperson of the Portuguese Association of Sanitary and Environmental Engineering between 1985 and 1991. He was a member of the National Water Council between 1994 and 1999 and again since 2005, which is the advisory body of the Portuguese government for national planning in the water sectors. He has been a member of the Centre of Studies in Public Law and Regulation at the University of Coimbra since 2003, and a member of its governing council since 2009.

He is actively involved at the international level, particularly through the International Water Association (IWA), the activities of which he started participating in 1982, when it was called the International Water Supply Association (IWSA). He was technical secretary of the respective Portuguese National Committee between 1992 and 1999 and chairperson of the Distribution Division of IWSA between 1994 and 1998. He has been member of the IWA Strategic Council and IWA Fellow and is member of its Board of Directors since 2012.

He was a founder of *Engidro, Estudos de Engenharia, Lda.*, a consulting company in water and waste engineering, for which he worked between 1979 and 1991. He was a founder of *Ambitec, Tecnologias para o Ambiente e Saneamento Básico, Lda.*, a company providing services in the field of support technologies for systems operations, for which he worked between 1986 and 1991. He was a founder of *Engiform, Engenharia e Formação, Lda*, a company providing technical training services.

He established the quarterly magazine *Ambiente 21*, and was technical director between 2001 and 2003. This magazine was aimed at the technical sector and the general public, dealing with environmental issues and their interaction with agriculture, industry, energy, transportation, tourism and leisure, urban development, sea and trade. Nine issues were published under his directorship.

He was honoured in 2012 with the IWA Award for Outstanding Contribution to Water Management and Science, which was awarded at the IWA World Water Congress in Busan, South Korea, for significant and innovative contributions with an international impact, related to leadership, management, funding, systems operation, technical innovation and research into water sectors activities.

In 2003 he became the chairperson of the Institute for the Regulation of Water and Solid Waste (IRAR), the regulator for these services in mainland Portugal. This entity was renamed the Water and Waste Services Regulation Authority (ERSAR) in 2009, which is his current place of work. ERSAR is an independent administrative authority, with about 70 employees. Its mission is to regulate the public drinking water supply, waste water management and solid waste management services in Portugal and it has the role of national authority for drinking water quality, regulating about 500 utilities that provide services to around ten million inhabitants.

Chapter 1
Introduction

The aim of this book is to present and propose an integrated regulatory approach for public services concerning three sectors, drinking water supply, waste water management and solid waste management, otherwise known as water and waste services. This approach shall be able to provide the almost total universal access of citizens to these services, supplied with suitable quality by the utilities, at socially acceptable prices and with an acceptable level of risk.

These services are essential to the well-being of citizens and the economic activities, and have a strong effect on the improvement of public health and environment, as well as economic benefits, thus contributing to a more developed and healthier society. It is well known that enough and safe drinking water is necessary to human being not only for survival, but also to remain hydrated and be able to achieve all his mental and physical potential.

Clarifying the terminology used in this book, drinking water, also written as potable water, is water pretending to be safe enough to be consumed by humans or used with low risk of immediate or long term harm. More precisely, it is all water intended for drinking, cooking, food preparation, personal hygiene and other domestic purposes, regardless of its origin and which is to be provided from a distribution network, lorry or tanker, or in bottles or other containers, whether for or not for commercial purposes. It is also all the water used in a food industry company for the manufacture, processing, preservation or marketing of products or substances intended for human consumption, as well as that used in cleaning surfaces, objects and materials that may be in contact with food, except when the use of such water does not affect the wholesomeness of the foodstuff in its finished form.

Waste water, also written as wastewater or sewage, is any water that has changed its natural characteristics and has been adversely affected in quality by anthropogenic influence, which means human use. Frequently waste water mixes with storm water, also written as stormwater, which is water resulting from the meteorological phenomenon of rainfall in the catchment areas of urban areas.

Solid waste, also written as municipal or urban waste, commonly known as trash, garbage, refuse or rubbish, is defined as any substances or objects which the holder discards or intends or is required to discard.

These services are generally provided, given the technologies currently available, under a natural or legal monopoly, so there is no clear incentive for utilities to search for greater efficiency and effectiveness. This raises the risk of the prevalence of such situation among services provided to the users. These services are really considered the last of the major monopolies. Society can significantly benefit from the existence of regulatory intervention, for these reasons. This intervention may act as a third party able to introduce greater balance in the relationship between the service utilities and their users, as well as greater efficiency and effectiveness.

A general trend in many countries in the regulation field in the last two decades, following the American tradition which has existed for more than a century, is to deliver regulatory tasks to public entities independent of governments and the direct administration of the State, designated as independent regulatory authorities. These are characterised by the following essential features: stable mandate for the independent regulatory authorities, which cannot be removed before the end of the contracted term, except in cases of serious misconduct; autonomy in exercising their regulatory functions, as they are not bound to specific orders, instructions, directives or guidelines from governments; the definitive nature of their decisions, which can only be challenged in the courts and cannot be reviewed by governments; substantial administrative autonomy in their human and budgetary resource management.

Regulatory intervention may be embodied in various forms, in terms of integrating the various components of its approach, universality of its application and intervention intensity, which should be in accordance with the real situation on the ground and of course the current policy option. It may range from implicit regulation provided by the State itself, as has always historically happened, for example through directorate-generals, to explicit regulation through a specialised agency, with greater or lesser powers delegated in it by the State and with a varying degrees of independence. Several models are therefore possible, though of course with different levels of effectiveness.

What is presented here is not an allegedly universal regulatory model, because this does not exist, but a model inspired on what was designed and implemented in the Portuguese context in the transition from the twentieth to the twenty-first century. This was a remarkable period in the development of water and waste services in Portugal, and the model presented here can in some cases be adopted or adapted to other situations and contexts.

The approach taken in this book does not focus on the theory of regulation, as that is the subject of many good publications, but on the practice of regulation, considering the experience of around twelve years of the Portuguese water and waste services regulatory authority.

The author argues, based on this experience, for a methodology that adopts a model of regulation based on an integrated approach (known in the abbreviated form as the model RITA-ERSAR) for water and waste services. In effect, it regulates both the sectors as a whole and the individual utilities, aggregates the various strategic, technical, economic, environmental and social components, seeking an appropriate balance of the various perspectives at stake. It is simultaneously applied to three different public services, drinking water supply, waste water management and solid waste management, with suitable adaptations, trying to find the optimal global solution.

This model generally follows a set of attributes that are considered important for regulatory intervention, insofar as this intervention must tend towards being integrated, topic-based, universal, geographically extended, flexible and cooperative. Each one of these attributes is analysed below:

Integrated regulatory intervention

What is desirable is a considerably integrated regulatory intervention, expressed in a regulatory model that considers both the structural regulation of the sectors as a whole, in order to develop its organisation, legislation, information and general training, and behavioural regulation of each of the utilities to do with their legal and contractual, and economic aspects, as well as quality of service, quality of drinking water and interface with users.

This holistic vision makes it then possible to take advantage of potential synergies between these regulatory components and promote better performance in the sectors and the utilities, as is explained in detail below.

Topic-based regulatory intervention

Regulatory intervention must be well defined in terms of the activities covered, that is to say the services covered by the regulation should be well identified. There are of course several possible solutions, where the same regulatory authority can cover a wide range of services or a single service.

A solution that is advocated here is one of relatively limited regulatory intervention, involving drinking water supply, waste water management and solid waste management, the three irreplaceable structural services for modern societies. Their proximity and interdependence are grounds for joint regulation, without losing any necessary specialisation. The regulation may well benefit from the potential of harnessing the efficiencies of closing the circle of the urban water cycle, by jointly regulating drinking water supply and waste water management, and the combined management of by-products, jointly regulating waste water management and solid waste management.

Universal regulatory intervention

Regulatory intervention should be universal, covering all water and waste services management models, whether State or municipal, public or private, to counter a relatively common belief that explicit regulation is linked to private management. In reality, regulation is just as important in a public as in a private sector context.

This ensures the same level of protection to the user, regardless of the type of utility providing the service, in terms of access, quality and price, which of course is very important in terms of consumer protection. Regulation may also be more effective when there is a wider field of comparison between utilities, between types of entities and between management models.

Geographically extended regulatory intervention

Regulatory intervention should be geographically extended, preferably at the national level, although this may be regional in large countries.

This enables a more global regulatory view of the sectors and the harmonisation of rules, procedures and interpretations in an extended territory, with the possibility of benchmarking a more significant number of utilities, reducing the risk of regulatory capture through the multiplicity of relationships and, of course, the rationalisation of regulatory resources, providing lower costs for users.

Geographically localised regulation focusing, for example, on urban areas only, in addition to the difficulties of operating efficiently and effectively for the reasons set out above, may even contribute to increasing inequality between urban and rural zones.

Flexible regulatory intervention

Regulatory intervention should be flexible, with a level of implementation strength that is appropriate in space and time to the reality of the sectors and the stage of development of the country, as well as the resources and capabilities of the regulatory authority proper.

Methodologies for direct regulation may for instance be adopted by intervening directly in the services, or regulation may be indirect, by influencing those services only through its power of influence, with binding or only advisory powers, according to the moment and the specific situation.

Natural development and the increasing sophistication of water and waste services entail the adoption of a regulation model that should tend to evolve over time.

Collaborative regulatory intervention

There are of course many ways to carry out regulation, and the model chosen in each situation should depend on the stage of development of the sectors and of course the sociocultural context.

It is suggested in this model, nonetheless, that the regulatory approach will mainly be collaborative, primarily using the power to influence. The power to sanction will only be used secondarily.

Hence, regulation should be an instrument supporting the development of the sectors and its stakeholders, and not just an instrument for supervision and control, as its ability to influence and empower stakeholders in the sectors, especially utilities, is enhanced by its overall and extensive geographical intervention.

Regulation must therefore adopt a strategic didactic posture in relation to the utilities, especially when they show weaknesses in the management of services, and efforts should not be spared to support their capacity building. An important aspect of this didactic posture is linked to the recommendations that can be made, either in general terms or for example following inspections. These can focus on aspects which, whilst constituting or not any infringement, are open to improvement.

The following chapters will briefly describe water and waste services, their respective public policies, the setting up of a regulatory authority, the integrated regulatory approach that is proposed, the respective components of regulation, coordination with third parties and finally the main conclusions.

Chapter 2

Water and waste services

2.1 INTRODUCTORY NOTE

Activities involving supplying water to populations and providing waste water and solid waste management services are essential public services of a structural character which are vital to the general welfare, public health and collective security of populations, their economic activities and protection of the environment.

These services are of fundamental importance in modern societies and are usually classified as being of general economic interest. Indeed, they contribute significantly to the social and economic development of the country both in terms of the increasing improvement they grant to the living conditions of the population and the environment, and their ability to generate economic activity and thus create employment and wealth.

They must therefore follow a set of principles, particularly universality of access, the ongoing provision and quality of service, efficiency and price equality, as explained below. These elements are an important factor of social balance. It is therefore not possible to speak of the real development of societies without taking into account their need to provide these services in a generalised and accessible manner.

The aim is to promote the provision of public water and waste services at an appropriate level of quality, at socially affordable prices and an acceptable level of risk. A further aim is to find the optimal long-term balance between these objectives, ensuring rationality and transparency to users, regardless of the ownership and management model of the services, whether at the State, municipal or private level.

This chapter presents the water and waste services, and considers public service obligations, the characteristics of the services, the sectors' stakeholders, the systems of the water and waste services, the link with the environment and water resources, the challenges for water and waste services, and also the rights of users of those services.

2.2 OBLIGATIONS FOR WATER AND WASTE PUBLIC SERVICES

Water and waste services must comply with a set of obligations for public services:

Universal access to services

These services, since they are essential and irreplaceable, should ideally benefit the entire population, through accessibility to public water supply networks, waste water and an solid waste collection system, whether door-to-door or in the vicinity. This is the way found to provide the population with services supplied by professional utilities, thus excusing citizens from having to provide these basic needs themselves with individual solutions.

It should be noted that in areas with a lower population density it is sometimes not economically feasible to set up traditional public systems, which would lead to unacceptable unit costs for the population. In these cases, the adoption of individual water supply systems is justified, such as holes or wells, and waste water treatment, such as septic tanks, as well as local disposal depots further from the places of generation and/or less frequent waste collection. More rural areas can, for example, be given a greater incentive for individual solutions involving composting of biodegradable solid waste.

Adequacy of services in terms of quantity

The services should be designed to provide an appropriate volume of drinking water, as well as deal with the volumes of waste water and the amount of solid waste produced. The existence of systems below capacity, which for example can only function intermittently or individual storage solutions, generally give rise to considerable health and environmental problems. Domiciliary drinking water reservoirs are potential causes of water quality degradation and the prolonged accumulation or uncontrolled disposal of waste resulting from inadequate collection frequency and/or container capacity has inevitable impacts on public health and the environment.

Adequacy of services in terms of quality

These services should be provided with quality, the concept of which is developed later in this book. They should ensure that, in addition to access, users have sustainable services in terms of their interface with the user, service management and environmental management.

In the case of water supply, for example, the suitability of the user interface or social sustainability encompasses access to the service and quality of service provided to users. Service management encompasses economic sustainability, infrastructural sustainability and human resources productivity of the utilities. Environmental sustainability encompasses the efficient use of environmental resources, namely water and energy, and the prevention of pollution by the utilities.

Continuity of services

These services must be permanently available to ensure an appropriate level of comfort to users, implementation of best practices in systems operation and greater protection for public health and the environment.

For this purpose, there must be sufficient water resources for water supply and the infrastructure must be appropriate to the need to provide continuity of service with adequate pressure maintained in distribution systems. In waste water management, functionally inseparable from drinking water supply and solid waste management, there must be infrastructure appropriate to the need to provide continuity of service.

There are, however, situations where, due to limitations of water resources or infrastructure, the service is intermittent. This situation is, naturally, to be avoided, because it has been shown to be ineffective, often require not less but more source water, and pose additional risks to public health.

Structural efficiency of services

These services must be efficient in structural terms, taking advantage of economies of scale, of scope and process. The sectors and utilities must be organised in an optimised manner, namely suitably sized at territorial viewpoint, and also with a suitable combination of services and ensuring the complete production process. Care should be taken that costs do not include components resulting from relevant structural inefficiencies, such as those arising from utilities which are far too small, as this often leads to higher tariffs for the user.

Operational efficiency of services

These services must be efficient in operational terms, in how they are managed, using resources available for optimised service provision by the utility. Care should be taken that costs do not include components resulting from operational inefficiencies, such as water losses, excessive energy consumption or excessive staffing of the utility, as this often leads to higher tariffs for the user, as well as unacceptable environmental impacts.

Adequacy of services pricing

These services entail investment and operational high costs that should naturally be recovered through revenues. These revenues can be obtained through the choice of charging tariffs, through the use of national, regional or local taxes, or even transfers from abroad, for example development support funds, which means taxes from other countries. However, in order to environmentally protect water resources and to be fair, these should be borne by users in the form of tariffs, making sure prices are as appropriate as possible, that is not insufficient or excessive, in order to simultaneously ensure the economic and financial sustainability of the utility and macro-economic accessibility to the service by users.

Fair prices for the services

It should also be ensured with these services that the tariff structure is the most appropriate, particularly as regards the consumption and environmental behaviour of each user. The charges should also consider the economic capacity of users in order to ensure greater justice in the distribution of the charges among them, as well as accessibility to the service by the economically weakest users.

Adoption of codes of good practice

These services imply the adoption of codes of good practice, such as the transparency of its management, of the tariff system and of funding, the involvement of various stakeholders and users of the services in the main decisions, the mechanisms to resolve conflicts and information access.

All these aspects will be discussed in more detail throughout the text.

The first four obligations, namely universal access to services, the adequacy of services in terms of quantity, the adequacy of services in terms of quality, and service continuity are strongly linked to the need for effective services, that is, to meet the objectives defined by society.

The remaining five obligations, namely the structural efficiency of the services, the operational efficiency of the services, the adequacy of services pricing, fair prices for the services and the adoption of codes of good practice, are closely linked to the need for efficient services, that is, the use of available resources for optimised provision.

It is these obligations together that have tended to promote universal access by users to these essential services, provided at an appropriate quality by utilities, at socially acceptable prices and with an acceptable level of risk.

In this framework, one of the tasks of the regulatory authority is to ensure compliance with the public service obligations imposed by law or contracted with the utilities, whether these are public or private.

2.3 CHARACTERISTICS OF THE WATER AND WASTE SERVICES

The water and waste services have important characteristics, some of which differentiate them from other services, and which should be borne in mind for the purposes of regulatory intervention.

They are irreplaceable services which deal with heterogeneous products, allowing potential economies of scale in terms of range and process; they tend to be regional, using assets designed for peak demand, with high-value, long-lasting and high immobilisation situations; they have a long period for return on capital and low elasticity between demand and price; and lastly, and certainly most importantly, they are natural or legal monopolies.

Each one of these characteristics will now be analysed in more detail:

Irreplaceable services

These services are irreplaceable, because the drinking water supply, unlike other services such as those supplying electricity, gas or telecommunications, has no viable alternative. Users cannot do without these services because in most situations they do not have any other viable option.

Unlike the public drinking water supply, which involves the delivery of a product to the user, waste water management and solid waste management entail the removal of products generated by the users, for reasons of comfort, public and environmental health, and as such one cannot consider their replaceability.

Services with heterogeneous products

These are services that deal with heterogeneous products in which the properties of drinking water for public supply, waste water and solid waste vary in space and time. For example, the quality of drinking water supplied varies from place to place and from day to day, making the management of the service more complex of course, and involving intensive monitoring to ensure continuous drinkability. The same applies to waste water and solid waste, the qualities of which also vary from place to place and from day to day, requiring careful management of the services so as to ensure environmental protection.

Services with potential economies of scale

These services enable potential economies of scale, in which the unit costs for installation and production tend to decrease, up to certain limits, for increasing quantities of demand. This means that it is important to determine the appropriate size of each utility, thereby increasing its structural efficiency. This also allows to attract higher quality management, and to afford more competent technical scientific support. This is of course limited by the organisation of these sectors as provided for in the current public policy.

These are therefore services that generally find their ideal size at the regional level, and not at local level, due to a lack of economies of scale, or at the national or the transnational level, due to the very high transportation costs. Administrative division and land occupancy often hamper the technical and economic management of these services, for example when there is a large number of very small and often fragile systems, without the scale to ensure efficient service provision.

Services with potential economies of scope

There are also potential economies of scope, where unit costs tend to decrease with diversification into activities with similar characteristics, for example jointly managing drinking water supply and waste water, or associating other complementary or subsidiary activities to these, provided that issues relating to free competition are safeguarded, such as energy production or the recovery of by-products, due to potential synergies arising from human resources, equipment and facilities. This means that it is important for each utility to find a suitable combination of services to provide, thereby increasing its structural efficiency.

It should be noted that for water and waste services in general there is not much potential for economies of scope due to a joint managing, since they are significantly different activities.

Services with potential process savings

There are also potential process savings, by adapting the vertical integration of services, with the utilities developing their activities at the various production stages needed to transform raw materials into goods and specified services, from abstraction to distribution in drinking water and from waste water collection to discharge in receiving bodies. The need for vertical integration from collection to final disposal for solid waste management services is however questionable, due to the different technologies adopted in the retail and bulk activities.

This means that, when each utility is carrying out the complete production process in the provision of water services, it is possible to reduce the respective unit costs, thereby increasing structural efficiency.

It should be noted however that the demand for increased competition in these sectors may lead to a choice involving vertical disintegration, as is usual for instance in the energy sector.

Services with assets designed for peak situations

These services are characterised by being normally designed and engineered for peak situations, for a distantly projected horizon and the period of the year with peak demand, thus yielding significant idle capacity that is not used most of the time. This idle capacity naturally entails additional costs and the provision of incentives to minimize this, for example through temporary use for other purposes, is important for tariff containment.

Services with high value assets

These services are characterised by using extremely high value assets, to the extent that they are a capital-intensive sectors, one of the most intensive of all public services. This implies the availability of a high level of financing for the construction of expensive infrastructures. This is true not only for drinking water supply and waste water management services but also for the solid waste management service, especially in relation to its treatment and recovery.

Services with long-lasting assets

These services are characterised by the use of long-lasting assets, built for extended lifetimes, usually many decades, and even more than a century in the case of abstraction and drinking water distribution and waste water collection networks. These assets, since they are essentially major civil construction works, largely underground, with a high level of difficulty to replace and involving specialised technology with significant costs associated, are usually designed with the long term in mind and normally there is no great incentive for rapid innovation.

Services with high immobilisation assets

These services are characterised by using infrastructure that involves high immobilisation, as they are dedicated to one specific purpose, the provision of that service. Consequently, it is very difficult or even impossible to sell or transfer them. As such, the respective investments can be considered irrecoverable, commonly referred to as sunk costs, which means that the infrastructure tends to be used until the end of its useful life for that specific purpose. For this reason too, there is normally no incentive for rapid innovation.

Services with long-term invested capital recovery

These services are characterised by a long period of invested capital recovery as a result of a strong relationship between the considerably high value and long-term nature of the assets with revenues stemming from low unit value tariffs, which generally make these essential public services affordable to users. As a result, invested capital recovery is only possible after several decades.

Services with low elasticity between price and demand

These services are characterised by low elasticity between price and demand, since they are structural services for basic needs. Indeed, an increase in tariffs has a limited effect on demand, in so far as users have no viable alternative services and therefore cannot live without them. Notwithstanding the fact that the tariff can be used as an economic incentive to encourage suitable consumption and environmental behaviour, its effectiveness is necessarily limited.

Services that form natural or legal monopolies

These services are natural or legal monopolies, to the extent that there is, for technological or legal reasons, a single utility to provide these services in each geographical area served. This creates restrictions on competition between utilities and, in particular, natural barriers to competition in the sectors and the entry of new utilities into the market. This is not, of course, a motivator for continuous improvement in management efficiency. Moreover, users cannot choose the utility that they want nor the price-quality relationship that they consider to be the most suitable.

In the cases of drinking water supply and waste water management, these services are natural monopolies and are probably the public service markets where this feature is most prominent. Natural monopolies arise when the cost structure is determined by a fall in marginal and average production costs as the size of the production system increases, due to economies of scale. In these cases the total production costs for a given aspect of demand are lower when there is only one service provider.

However, in terms of market structure, the solid waste management sector is distinct from the water services sectors. There is no typical natural monopoly in the collection and transportation of solid waste because this is not a networked industry, and the services are provided often under a legal monopoly. In the case of waste storage, sorting, recovery and disposal, these can once again be characterised as natural monopolies.

Knowledge of the main characteristics of water and waste services is crucial for the specification of a regulatory model, but it is also important to identify the main stakeholders in the sectors, which will be described in the next item.

2.4 STAKEHOLDERS IN THE WATER AND WASTE SERVICES SECTORS

These sectors contains numerous and diverse types of stakeholders, which it is important to be aware of and briefly characterise. They can be divided into several groups, such as public administration, service holders, utilities, other entities providing services and civil society.

Each of these stakeholder groups will now be analysed in greater detail:

Public administration

The public administration is responsible for pursuing the strategies for the water and waste sectors at the national, regional and local levels. General mention should be made here of the importance of the environmental authority, the water resources authority, the waste authority, the health authority, the consumer protection authority and the competition authority, in addition of course to the water and waste services regulatory authority. These functions can of course be exercised separately or jointly by various bodies, which can vary from country to country, depending on the institutional organisation of each.

Services' holders

The services' holders have, in accordance with the law, the responsibility for ensuring the provision of the water and waste services, directly or indirectly, whether or not responsible for their management. This of course depends on the institutional structure of each country, but these usually involve central government, regional government and municipal councils.

Utilities

Utilities are responsible for the management of the services, in direct relationship with the end users or other utilities, regardless of whether they are service holders. They should be empowered to act by the service holder, generally via a contract. This of course depends on each country's choice of public policy, but they are usually the service holders themselves, State-owned, regional or municipal companies, or private companies.

Other service providers

Other service providers are not responsible for the management of the services but bridge gaps in the actual capacity of the utilities, through providing specific expertise. Of note are the companies operating and building systems, manufacturers and suppliers of materials, equipment and products, consultancy and project supervision companies, information systems companies, inspection and audit companies, quality assurance companies, research and development centres, training centres, analytical and testing laboratories and, in the particular case of waste, the solid waste collection or container cleaning companies. The funding agencies, focused on financing the necessary investments, are also included here since this is a capital-intensive industry. Insurance companies also form part of this group, with regard to the risks associated with these services.

Users

Users are natural or legal persons, public or private, who are provided, on a continuous basis, with water and waste services. They can be classified as end-users or user utilities.

End users are natural or legal persons, public or private, who are provided, on a continuous basis, with a water and waste service, and whose purpose is not the provision of that service to third parties. On the other hand, the user utilities are public or private legal persons, who are provided, on a continuous basis, with a water and waste service and whose purpose is the provision of that same service to third parties.

End users can be classified as domestic users, who use urban buildings for residential purposes, or classified as non-domestic users, which encompasses all others, including the State and municipalities.

Civil society

Civil society is involved in these sectors through the user's protection associations, the economic and business associations, environmental protection associations and the mass media. This also refers to the general public, who may not necessarily have a contractual relationship with the utility, but can be affected by the service it provides.

The water and waste services utilities are stakeholders particularly important for regulation and are characterised by:

- Being organisations that have human, technical and financial resources and important physical infrastructures and equipment in order to meet the service needs of their users.
- Using environmental resources for their activity, mainly surface water and groundwater resources, and, conversely, cause impacts resulting from their activity that may be relevant for the environment, particularly in relation to waste water and solid waste management. They may also be producers of raw materials and energy, for example resulting from processes of recycling and recovery of solid waste, avoiding the use of natural resources.

An entity must cumulatively meet the following conditions for it to be considered an utility legally permitted to provide water and waste services:

- Manage wholly or partly at least one of the services of drinking water supply, waste water management and solid waste management.
- Operate within a public system providing the aforementioned services, actually undertaking a current activity, which is clearly defined geographically, having a population that is served or potentially served by it and be able to identify and characterise its users.
- Be part of one of the management models provided for in legislation, for instance direct management, delegated management or concession management.
- In the event it is not directly managed by the holder, holding a legal instrument issued by the holder, for example a contract which formalises their responsibilities in providing the service, continuing in this way to be responsible for the service even when subcontracting other operators to carry out some of its duties.

2.5 NECESSARY SYSTEMS FOR WATER AND WASTE SERVICES

Water and waste systems are functionally interconnected sets of legal relationships that comprise the necessary support for the provision of these services by the utility and include rights and obligations embodied therein:

- Public service obligations, that is to say the utilities undertake to develop all the necessary and suitable activities for correct service provision.
- Legal relationships with the service holder, namely contractual obligations, for example with the delegator or grantor, in cases where management is not direct.
- Legal relationships with users, namely contractual obligations, for example with the users concerning supply or collection.
- Labour-related legal relationships, namely contractual obligations with their human resources.
- In cases where management is direct, legal relationships concerning ownership of infrastructure, facilities, basic transportation and administrative equipment, tools and utensils, consumable and replaceable stocks.
- In cases where management is not direct, such as delegations and concessions, legal relationships determining use of infrastructure, namely contractual obligations, free of charge or with a cost, for example with the delegator or grantor.
- Legal relationships with suppliers, namely contractual obligations with third parties relating to the acquisition of goods or services.

The water and waste infrastructures are described below, given the importance of its role among these system components.

The infrastructures, or assets, are items of a long-lasting or permanent nature in utilities, controlled by them and in terms of what is expected they form a future economic benefit within the scope of regulated activity. They are not intended for sale or transformation in the course of these activities as utilities.

A drinking water supply system can normally be considered to have various infrastructure components, from the water source to the point of consumption, in general the tap. These are, in most cases, as follows (Figure 2.1):

- Abstraction system, with the function of collecting, at any surface or underground water resource, untreated water in sufficient quantity and with minimally acceptable physical, chemical and bacteriological properties.
- Treatment system, with the purpose of correcting the physical, chemical and bacteriological properties of the untreated water in order to make them compatible with the requirements established for drinking water.
- Pumping system, which has the function of transmitting energy to the water, so that it can deal with uneven topographies and circulate at a suitable pressure, ensuring the necessary conditions at points of consumption.
- Transport system, which has the function of conveying water, usually without service connections, from the abstraction area to the consumer zones, sometimes far away.
- Storage system, which has the function of storing water for varying periods, primarily to stabilise flow, but also to ensure reserves for fire or service failure.
- Distribution system, which has the function of distributing within the supply area, based on a network, ensuring suitable pressures and flow at the various consumption points.
- By-products recovering system, to take advantage of and recover by-products such as sludge and energy for uses according to its characteristics and quality.

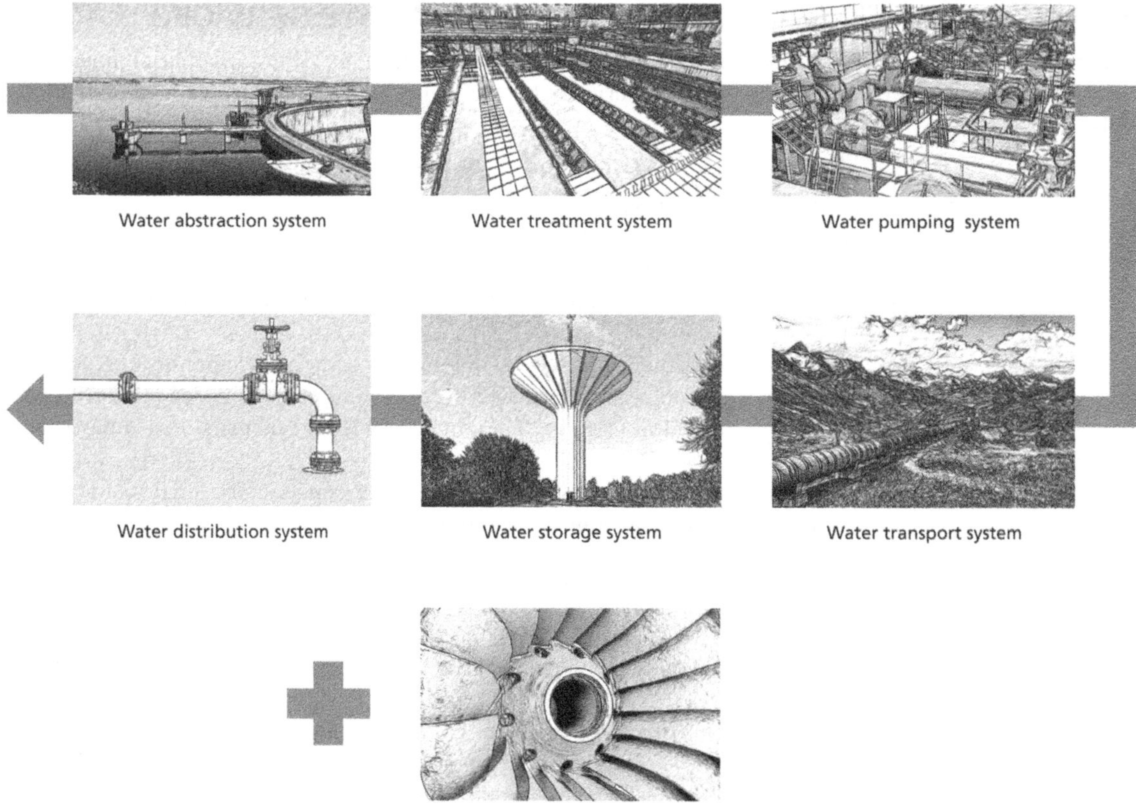

Figure 2.1 Drinking water supply infrastructures.

A waste water management system is normally made up of various infrastructure components, ranging from production to the disposal point. These are, in most cases, as follows (Figure 2.2):

- Collection system, with the function of carrying out the waste water and taking it to the appropriate destination.
- Pumping system, which has the function of transmitting energy to the waste water so that it can deal with uneven topographies.
- Transportation system, which has the function of transporting waste water, usually without service connections, to the treatment area, sometimes far away.
- Treatment system, which has the function of reducing the pollution load of the waste water, in terms of its physical, chemical and bacteriological aspects, in accordance with the features and consequent dilution and assimilation capabilities of the receiving environment.
- Discharge system for treated waste water, which has the function of returning the suitably treated waste water into the receiving environment.
- Sludge treatment system, which has the function of suitably handling the sludge resulting from treatment, known as its solid stage.

- By-product reuse and recovery system, such as waste water and sludge, suitable for uses according to its characteristics and quality, for example for irrigation and energy production.

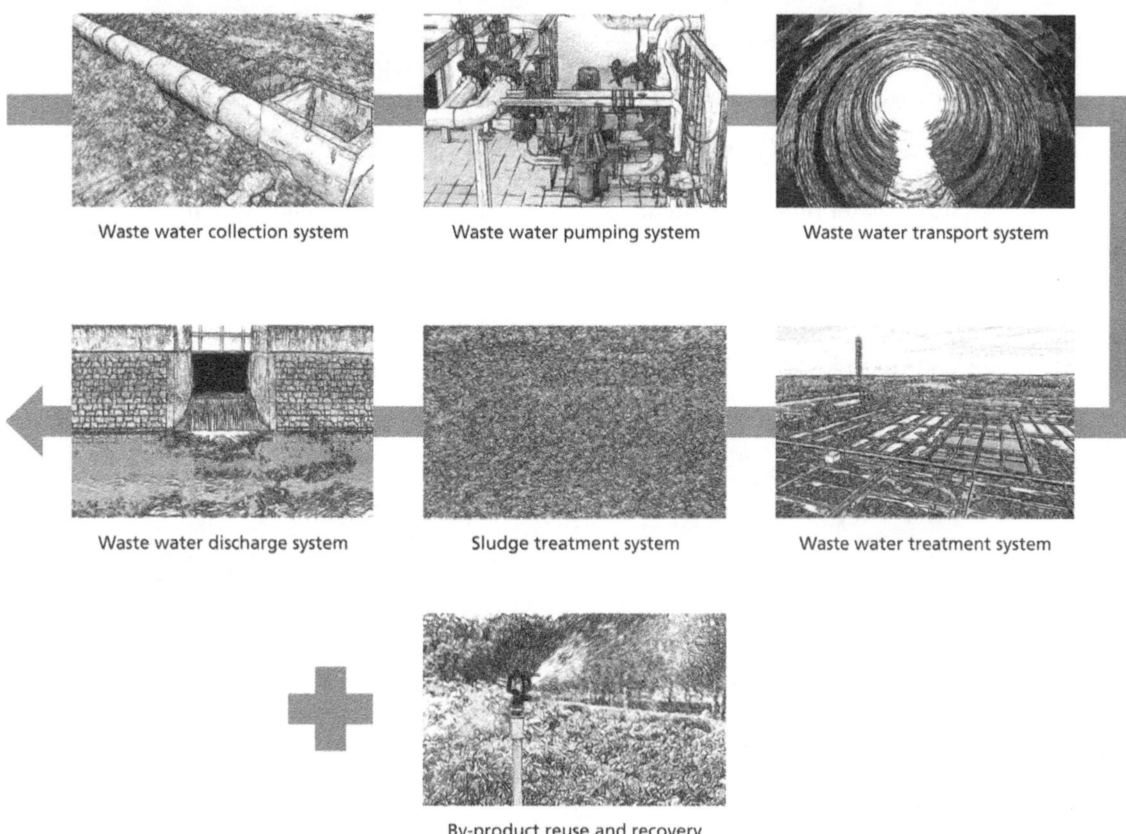

Figure 2.2 Waste water management infrastructures.

Strongly linked to waste water services, storm water drainage is a recurrent occurrence, and the two systems may be physically separate, joined or mixed. In the first case, a storm water system can be considered to have various infrastructure components, ranging from where those waters are generated to the discharge location. These are, in most cases, as follows (Figure 2.3):

- Drainage system, which has the function of collecting the storm water and taking it to an appropriate destination.
- Retention system, which has the function of reducing the flow peaks of storm water to avoid urban flooding and overloading the drainage and treatment systems.
- Treatment system, sometimes used with the function of reducing the pollutant load of storm water, in terms of its physical and chemical aspects, in accordance with the purifying capabilities of the receiving environment.
- Discharge system, which has the function of placing the storm water in the receiving environment, suitably treated where necessary.

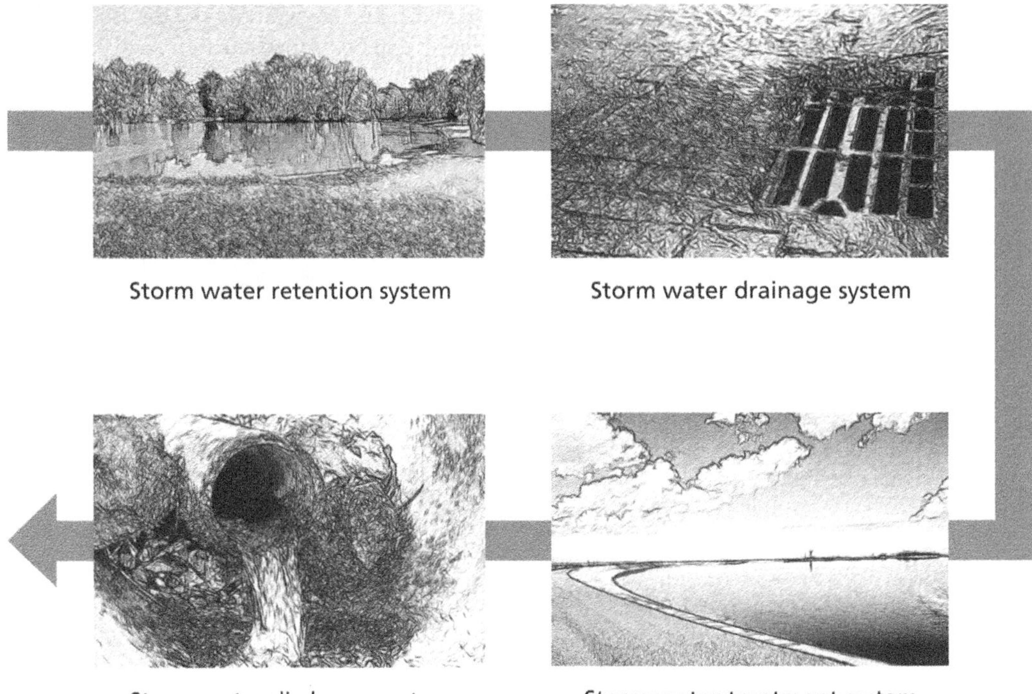

Figure 2.3 Storm water management infrastructures.

A solid waste management system can normally be considered to have various infrastructure components and equipment, from its collection point to its recovery or disposal point. These are, in most cases, as follows (Figure 2.4):

- Collection system, which has the function of collecting the solid waste for purposes of transport to a waste treatment facility; it is divided into undifferentiated collection, which consists of solid waste collection without prior selection, and selective collection, which is collection made to maintain the flow of separated solid waste according to type and nature in order to facilitate specific treatment or recovery.
- Transportation system, which has the function of transporting the solid waste, in general by trucks, in an undifferentiated or selective manner.
- Transfer system, which have the function of concentrating the solid waste collected for later transportation to the treatment unit.
- Storage system, which has the function of storing solid waste for varying periods to regulate its flow.
- Sorting system, with the function of selecting the solid waste to be recycled or for other forms of recovery.
- Recovery system, which has the function of making the solid waste treatment process profitable through its transformation to produce materials for recycling, including compost, electricity or heat.
- Waste disposal system, which has the essential function of channelling part of the solid waste arising from recovery operations, intended to be disposed of in a landfill.

Figure 2.4 Solid waste management infrastructures.

A regulatory model should also bear in mind that, notwithstanding the fact that a very large majority use the public systems described here, part of the population lives in small communities. It is thus not possible to reach global acceptable levels of service without paying particular attention to solutions using appropriate technologies for small communities, which are generally characterised by greater technical simplicity and therefore lower cost.

In fact, the so-called traditional solutions tend to lead to much higher costs per capita when applied to small communities, mainly due to the high transportation costs. Such solutions are difficult to implement, not only due to the high investments that are required by the utilities as well as the major operating costs that cannot easily be borne by users, who normally correspond to a population group of lower economic resources.

It thus becomes interesting to adopt more appropriate technical and economic solutions for these small communities, which can achieve basic satisfaction levels for the population with little use of complex technology, implying relatively moderate initial investments and easy operation, consequently resulting in lower charges.

A regulatory model should also take into account the life cycle of this infrastructure, which can be considered to consist of several stages, namely planning, design, construction, operation and finally rehabilitation.

Each of these stages of the infrastructure life cycle will now be analysed in more detail:

Planning

The beginning of the life cycle corresponds to planning the system, a stage in which it is conceived in general terms, taking into account firstly the physical environment in which the system will be located and, on the other hand, the objectives intended for it. Suitable physical infrastructure planning is an essential step in ensuring system quality as a whole and in particular its subsequent correct functioning. This planning must be undertaken at a national, regional and local level, ensuring suitable links with other strategies, particularly water resources, solid waste, land use and public health. The possible absence of this stage unavoidably has a strong negative impact on the optimisation of the expected results from the investments made. This can result in non-optimised infrastructure without the benefit of economies of scale, often disjointed within itself and with neighbouring utilities. This results in subsequent difficulty in interconnection, with a possibly large number of dispersed small-scale infrastructural elements, posing problems of increased construction and operational costs. There must therefore be master plans prepared by the utilities, in order to ensure a global vision of the works to be undertaken incrementally over time, allowing regionalised solutions that are rational and economically optimised.

Design

This is a stage to draw up the project, which involves designing, detailing and specifying the infrastructure planned in the previous stage. Suitable design and a correct specification constitute an essential step in ensuring overall quality, especially for subsequent suitable operation of this infrastructure, as well as optimising the use of the different facilities.

Construction

The construction of the infrastructure is then implemented, allowing it to be placed at the service of the users with a useful life intended for as long as possible, in general several decades. An essential stage of the process is suitable technological resources being used in the construction process along with the appropriate use of materials and mechanical and electrical equipment, to ensure overall quality and in particular as regards the future appropriate functioning of this infrastructure.

At this stage, another essential step is appropriate control and supervision of the construction works to ensure overall quality and, in relation to the future, the suitable functioning of this infrastructure. Any errors or omissions during this stage will necessarily have a strong negative impact throughout the useful life of the infrastructure, corresponding to premature deterioration and lower reliability, frequently causing higher operating costs and even expensive renovation works.

Operation

It is then possible to start the operational use of the infrastructure, with the corresponding utilisation and maintenance. Suitable utilisation is an essential activity for ensuring the quality of service to users. All utilities need to strive for a high level of efficiency and effectiveness in their operations in order to achieve their management objectives. Efficiency measures the extent to which available resources are used in an optimised manner for service creation, while effectiveness measures the extent to which defined management objectives are met.

Rehabilitation

If no other intervention is carried out, the natural and inevitable aging of the infrastructure and/or the possibly of its accelerated aging, resulting from errors in planning, design, construction or operation, gradually tend to hinder and even make it impossible to meet its objectives in a technically and economically acceptable manner, leading to the end of the life cycle of the infrastructure.

It is however possible to use rehabilitation, framed within an appropriate asset management strategy, with the goal of improving performance by changing its physical state and/or technical specifications, enabling these elements to keep operating for a significantly longer period of time.

The greatest needs in developed countries are increasingly focused on the rehabilitation of infrastructure, mostly underground in the case of water services, with a strong impact on populations, such that asset management has gradually reached a core level of importance.

Indeed, the low levels of infrastructure rehabilitation prevailing in much of the world constitute a serious threat to the long-term sustainability of water and waste services. Rehabilitation is often imperceptible to society and to policymakers, and thus transferring an excessive burden to generations to come. A paradigm shift in asset management is therefore urgently needed, given the restrictive context of provision of capital and increasingly stringent environmental constraints.

2.6 LINKING THE SERVICES WITH WATER RESOURCES AND THE ENVIRONMENT

The water services, defined as the drinking water supply and waste water management, components of the urban water cycle, are strongly dependent on water resources, as these constitute the source for the production of drinking water and are also, in most situations, the final destination for waste water. Suitably treated waste water is generally disposed of in water resources and, after extensive dilution and natural purification, reused indirectly through the water cycle, after appropriate treatment. It is also directly used for other purposes, such as agriculture.

There is therefore a strong interaction between water services and water resources, as shown mainly through two linking points, namely water abstraction and the discharge of waste water.

On the other hand, solid waste services are highly dependent on the environment, as this is the raw material base for the manufacture of consumer goods which, after use, give rise to solid waste and which also, in most situations, constitute its final destination. Indeed, solid waste is disposed of in the environment and subsequently reused indirectly in the long run through the natural life cycle of materials or, more directly and in the short term, reused for various purposes as a raw material or resource after treatment or recycling. That which cannot be reused or recycled is subject to energy recovery, despite the emission of pollutants into the atmosphere and the production of fly ash and slag. Protection from the impacts is required. When deposited in landfill leachate may be produced, the impacts of which also have to be prevented and safeguarded.

There is therefore a strong interaction between the solid waste services and the environment, which is primarily embodied in one linking point, the disposal of solid waste. Actually, the mining of materials is obviously not a public service but a commercial activity, although there can be a strong interaction between the mining activity and the use of solid waste as a raw material, since these can replace virgin materials, reducing the need for their mining.

There is also an interconnection between water services and waste services, mainly because the treatment of drinking water and waste water treatment gives rise to important volumes of sludge, which requires disposal. Energy can be extracted from that sludge, which can then be used in the water and waste services themselves.

These services are also strongly correlated with land use, and any urban planning options or omissions, resulting for instance in an excessive dispersion of housing, imply increased costs for water and waste services, tariff increases and the reduction of economic and/or physical accessibility to the service.

Especially in developing countries, with rapid urbanization, water and waste planning should be an integral part of city planning.

A particularly important aspect in the linking of water resources with water services is the efficient use of water. Indeed, water is a scarce resource and it has become necessary to increase the efficiency of its use. This is the case not only because it is an environmental imperative, but also due to the strategic need to preserve its availability and the water reserves of each country and the importance of water at the national level among the economic activity, utilities and citizens. Users should not have to pay for inefficiencies resulting from an excessive amount of water losses in public systems and they should be aware of the need to reduce their own water losses and water waste. It has the additional advantages of reducing pollution stemming from the resulting waste water and also associated energy consumption.

In public drinking water supply systems, the factors that contribute to these inefficiencies are typically water losses in networks and inefficient procedures for washing pavements and watering gardens and sports fields. In housing, it is the inadequate use of flush toilets, showers, taps, washing machines and dishwashers, washing vehicles, heating systems air conditioning and pools.

As far as the measures to be adopted in public drinking water supply systems, a reduction in water losses should be encouraged and also the use of treated waste water. Losses constitute one of the major sources of inefficiency and should therefore be the object of special attention and a control and minimisation strategy by the utilities. It is also necessary to define an effective national or regional strategy regarding the use of treated waste water, to meet the desirable objectives for the utilisation of this resource.

In dwellings, users should be encouraged to use water rationally and choose appliances and fittings such as toilets, showers, taps, washing machines and dishwashers, heating and air conditioning systems with low energy consumption. In terms of outdoor uses, it is necessary to promote suitable procedures for housing and the use of low energy consumption equipment for the washing of pavements and vehicles, watering gardens and sports fields and pool management.

The regulatory authority should play an important role together with the utilities and users to promote the efficient use of water, as well as the proper management of the solid waste generated.

2.7 CHALLENGES FOR WATER AND WASTE SERVICES

A prospective analysis of what might be the most important challenges for water and waste services within a medium and long-term perspective is presented below, in order to achieve a better understanding of these challenges by the regulatory authority and the timely preparation of its regulatory strategies to respond to them.

These challenges tend to exist primarily in terms of the management of public drinking water supply, waste water management and solid waste management services, but also in the service interfaces with users, regarding public health and the environment and in relation to the models of governance adopted.

In some cases they may not significantly affect the activity of the utilities, but in other cases gradual adjustments or even the rapid adoption of alternative routes may be required. This analysis considers the predicted evolution of various social, environmental, economic and technological factors and governance options for the sectors and their consequence for the services in terms of public health, the environment and even national security.

The following main challenges can be identified with regard to the management of the public drinking water supply, waste water management and solid waste management services:

* Water abstractions will be more at risk of degradation in their quality due to the effect of chemical and emerging biological pollutants. The intensification of indirect reuse of water in nature, along with a growing increase in abstractions and discharges, explains the growing potential presence of

emerging pollutants in drinking water. In recent decades, despite urban domestic pollution being strongly controlled through treatment plants for the removal of macro-pollutants such as organic matter, nitrates and phosphates, the amount of chemicals used in agriculture, in industry and even for domestic purposes, such as pesticides, solvents, and pharmaceutical residues, has increased enormously. New biological pollutants have also appeared as a result of biotechnologies. There will therefore have to be a growing effort to protect abstractions, mainly from diffuse and accidental pollution, and in particular through the definition and management of catch basin protection areas, and an increase in the use of water treatment solutions with multiple barriers and the implementation of safety plans for water and other approaches to management and risk assessment.

- Water abstractions will be more subject to the risk of decreased water availability in nature and an increase in seasonal and spatial asymmetry, as well as deterioration in quality, due to the effect of climate change. More frequent and extensive extreme hydrological phenomena such as hot and dry periods may occur in some regions in the medium term, if the predictions for climate change are confirmed. This situation increases the risk of a shortage of drinkable surface water and groundwater, both in terms of quantity and in terms of quality, compounded by the expected increase in consumption. Insufficient quantity can result from insufficient replenishment of water resources through precipitation, while deterioration in quality may result for example from an insufficient ability to reduce pollution discharged into water resources, a decrease in the dissolved oxygen, the increased appearance of toxic algae on surface water and saline intrusion into aquifers due to overuse or rising sea level.

- Water abstractions to drinking water supply will be more subject to the risk of growing competition from other uses of water, leading to possibly greater shortages of available water resources. Given the possible increasing scarcity of available water resources, there will be a tendency for the utilities providing water services to experience increased competition from other users consuming water, such as agriculture in order to grow food and industries for the manufacture of products, or even the increasing ecological requirements related to water resources. It is essential for the integrated management of water resources to be strengthened by environmental authorities, despite the consensual priority associated with the extraction of water for drinking supply.

- Demographic changes will alter the amounts of water and waste, and consequently affect the utilities and possibly their economic and financial sustainability. In certain regions, the aging population in rural areas, the reduction of the birth rate, increasing life expectancy and the reduction in the number of inhabitants per household are demographic changes that will alter consumption patterns and affect utilities in terms of the size of their operations and the workings of their infrastructure. This evolution, which partially depends on land use policies, can pose problems to utilities in rural areas, with the underutilisation of existing public drinking water supply, waste water management and solid waste management infrastructure, creating idle capacity and thus hampering the economic and financial sustainability of these entities. It can occasionally also cause problems to utilities with the overuse of existing infrastructure, for example in the vicinity of urban areas, creating the need for new investments. Elsewhere in the world, phenomena growing in the opposite direction can be seen due to demographic explosions and major impacts on land use, with the need for infrastructure to respond to the accelerated growth of populations.

- Utilities are under increasing pressure to offer greater efficiency in the provision of water and waste services to keep prices affordable. Utilities will increasingly be faced with the dilemma of the need to make significant investments in infrastructural development and the increased charges in their operation, in the maintenance and rehabilitation of systems, in a social environment hardly favourable to a significant increase in the prices of these services, and certainly with a significant percentage of economically disadvantaged users. These factors will tend to put utilities under increasing pressure to offer greater efficiency in the provision of these services, seeking to ensure their service at the appropriate level of quality, with minimal resources and waste. There is a notable increase in the tendency for public disclosure and assessment of the efficiency of utilities.

- Water and waste infrastructure elements will become increasingly subject to risks associated with their aging and consequent degradation, thus encouraging new procedures for asset management. Their extended useful life periods and the significant costs of maintaining this infrastructure, along with tariffs that often do not ensure the recovery of costs, increase the risks of infrastructure aging and therefore a lower quality of service, system breakdowns or even the collapse of services, which can be translated into for example interruptions of service supply and pipe breaks, wasting resources, increasing costs and damage to third parties. As a result, the utilities will increasingly implement new asset management procedures, in order to analyse the system in an integrated manner and prioritise maintenance and rehabilitation investments according to actual needs and available funds, seeking to ensure the continuity of their useful life and the provision of the service with an appropriate level of quality.

- The behaviour of users, faced with the need for a more efficient use of water, will change their consumption patterns and affect utilities and their sustainability. The pressure exerted by authorities and the growing environmental awareness of users regarding the forecast increasing scarcity of available water resources will naturally improve the efficiency of water use for urban purposes, particularly regarding housing. Similarly, it will also reduce the production of solid waste. This environmentally necessary development can, in some cases, pose problems to utilities in terms of the underutilisation of existing infrastructure, creating idle capacity and thus occasionally affecting the economic and financial sustainability of these entities. The solution to this is the use of idle capacity for other purposes and or a tariff correction.

- The management of water and waste infrastructure will promote a more efficient use of energy in relation to its rising cost and environmental limitations. Energy costs are a very significant part of the operating costs of water services, and the increasing complexity of systems will tend to worsen this situation. There will be a tendency for utilities to seek greater energy efficiency and to increasingly use renewable energies, in some cases becoming small energy producers, due to the cost of energy being likely to increase significantly, which is responsible for the emission of greenhouse gases into the atmosphere. Utilities can produce energy through the use of biogas from landfills or the anaerobic digestion at solid waste units, energy recovery from the solid waste itself, which can be very significant, the hydraulic pressure available in the drinking water mains, and the sludge produced in waste water treatment.

- The technological development in water services, as in many other areas, will combine nanotechnology, biotechnology and information and communication technologies and allow very significant improvements, particularly in terms of processes involving water treatment and the removal of pollutants, improved pipes for water transportation, more widespread infrastructure automation, better monitoring through new sensors, telemetry, and so on. Similarly, technological developments in solid waste services will combine and optimise mechanical, biological and biotechnological processes that will improve the recovery of various materials and compost and will enable energy production resulting from the treatment of solid waste. This development will include information and communication management technologies that will enable very significant improvements, particularly in terms of recovery processes and the treatment of solid waste, more widespread infrastructure automation, better monitoring through new sensors, telemetry and optimisation of systems for the disposal and collection of solid waste. This development, although it will probably continue to be slower than in other network industries, such as energy and telecommunications, could alter the characteristics of the sectors, eventually making more competition possible and therefore making it less of a monopoly.

- The infrastructure elements for the abstraction and treatment of surface water will be more at risk from river flooding, due to the effect of climate change, causing the deterioration of water quality and possible destruction of infrastructure. In some regions there may be more frequent and extensive extreme hydrological phenomena, such as increased rainfall and corresponding river flooding, in the medium term. This situation increases the risk of the temporary impossibility of abstracting surface water for drinking water, through deterioration in its quality and the possible damage or even destruction of abstraction infrastructure. There will be, among other measures, a greater tendency for the creation and utilisation of large reservoirs of raw water bulk of treatment plants.

- Water and waste infrastructures will become subject to risks associated with urban flooding, due to the effect of climate change. If predictions about climate change are confirmed, any increase in rainfall and urban flooding will imply the rethinking of the management of public spaces, the conception, design and operation of drainage infrastructures due to the larger storm water flows, especially with the increase of permeable areas, the introduction of detention basins and the interconnection of collecting systems. It also implies rethinking both the design and operation of waste water drainage and treatment infrastructure, due to the predicted increase in seepage flows and subsequent alteration of the characteristics of waste water to undergo treatment.

- The utilities will tend to strengthen the reliability of the water and waste services infrastructure, particularly through contingency plans for situations resulting from natural and man-made accidents. The indispensability of these services is not compatible with interruptions to their functioning and even less in situations of prolonged interruptions resulting from natural disasters, such as earthquakes or caused, for example, by pollution accidents. The utilities will therefore tend to strengthen the reliability of systems with more resilient structures, greater physical interconnection between systems, alternative reserve abstractions and strengthen the storage of raw and treated water, despite the additional costs that involves.

- The water and waste services infrastructures that have a massive impact on the population may become terrorist targets. Existing social tensions and the increasing globalisation of society, with a considerable movement of people and the rapid dissemination of information, tends to increase the risk of terrorist acts. The infrastructure elements of these services, with their potential impact on populations, may become potential targets. The utilities will have to pay more attention to these aspects and reinforce the reliability and security of their systems.

- The treatment of waste water will become increasingly complex for utilities, especially due to the constraints of environmental legislation, the presence of emerging chemical and biological pollutants and climate change. If the predictions for climate change are confirmed, in some regions there may be more frequent and extensive extreme hydrological phenomena in the medium term, with more management difficulties. This could be worsened by the fact that waste water will contain significant amounts of storm water and seepage, which will alter its characteristics and decrease the efficiency of its treatment. This lowering of efficiency will also tend to be aggravated by the rise of emerging chemical and biological pollutants. Environmental legislation will also tend to be more demanding, thus further complicating treatment.

- Emerging markets for the reuse of water and recovery of sludge will presumably growth. The increasing scarcity of water resources and restrictions on waste water discharges into receiving waters will tend to encourage the development of emerging markets associated with the use of treated waste water for other purposes, such as agriculture, irrigating gardens and golf courses.

Moreover, the restrictions on discharges of sludge into the environment and the rising cost of energy will tend to encourage the appearance of emerging markets associated with the use of such sludge, for example for the production of energy.

- Emerging markets for solid waste recycling will certainly rise. Increasing restrictions on the disposal of solid waste in landfills and the requirement to fulfil ever more ambitious solid waste recycling targets will tend to encourage the emergence of markets associated with the use of waste for other purposes, while new raw materials, which are of interest to the industry in general will, by replacing virgin raw materials, decrease the pressure on their mining. Moreover, the rising cost of energy will tend to encourage the emergence of markets linked to the use of solid waste for energy production.
- The discharge of storm water will become increasingly complex for utilities, especially due to constraints from environmental law and climate change. There may in the medium term be more frequent and extensive extreme hydrological phenomena, if the predictions for climate change are confirmed, causing more difficulties in the management of drainage systems. This is compounded by the fact that storm water contains significant amounts of pollutants, especially in the initial period of precipitation, through its superficial washing of the ground.

The following are the most important challenges that can be identified at the interface between services and users:

- Users of water and waste services, through themselves and through their representative associations, will become more demanding and sophisticated, with a tendency to move from user to client perspective. They will require more information and transparency from utilities and will want to be more involved in setting public policies, the quality of services and tariff policies.
- Utilities will increasingly improve the quality of services in general and drinking water services in particular, given the growing demands of society and legislation, by strengthening communication mechanisms with users to provide more accurate public perception in this regard. The utilities will continue their improvement efforts, particularly with regard to the quality of drinking water, checking an increasing number of parameters in accordance with stringent monitoring programs, and using increasingly sensitive measuring instruments. However, the complexity of the problem, the growing number of factors and the lack of knowledge of the long-term impacts on human health of small concentrations of certain emerging pollutants will increase risk for the utilities.
- Given the increasing demands of society and legislation, utilities will increasingly invest in infrastructure and have significant costs regarding operation, maintenance and rehabilitation of systems, using progressively complex technologies. However, the idea that these are essential public services, and as such taken for granted by the general population, increases the risk of poor user perception of the actual costs involved and the need to pay an appropriate price for these. Thus, utilities will increasingly have to strengthen their communication mechanisms, to provide a more accurate public perception regarding the real costs of these services and the need for appropriate tariffs, whilst safeguarding concerns regarding social accessibility.
- The utilities will increasingly take measures to minimize the negative aesthetic and landscape impacts caused by water and waste services, for example through more careful architectural and landscape approaches for infrastructures, their greater integration into the landscape, the use of buried facilities, or visible facilities but with adequate architectural studies, the use of traditional finishes of the visible public facilities and promoting visits to facilities to enable increased awareness.
- The utilities will increasingly take measures to minimize the impacts on the movement of vehicles and pedestrians caused by water and waste services, for example through better planning and articulation of their own interventions underground, joint interventions in the subsoil with other management

entities services, assessing the feasibility of technical galleries for housing infrastructure, the use of new technologies installation and repair without digging trenches, adequate reconstruction of the pavements after closing trenches, adequate prior notice to users of underground works and optimization of the disposal and collection of solid waste circuits.

- The utilities will increasingly take measures to minimize the impact on user comfort caused by water and waste services, for example through adequate specifications for construction in terms of acoustic comfort, the removal of certain types of infrastructure from urban areas, the selection of less noisy equipment regarding for instance pumping stations when installed on public roads or in urban areas and the installation of odour abatement devices.
- Street works carried out by the water and waste service utilities, involving the maintenance, repair or rehabilitation of infrastructure, will have increasing associated social costs, due to their inconvenience and the nuisance they cause to citizens and commerce in general, especially in urban centres. They will be part of the economic analyses and limit the technologies to be utilised. It is thus envisaged that there will be increasing use of more sophisticated technologies, such as remote control, which although currently have higher costs than traditional methods, imply lower social costs. There will be also greater competition with other providers of public services in the use of soil and subsoil in urban areas and there will be increased regulatory constraints on their use.

The following most important challenges can be identified as being at the interface of services involving public health:

- The risks to public health linked to poor provision of water and waste services can increase, particularly as a result of demographic change, climate change and new chemical and biological pollutants. Firstly demographic changes and reducing consumption may accelerate the degradation of water quality in the pipes, due to longer retention times, or impair treatment efficiency. Secondly, if it is confirmed that some regions will have hotter and drier periods due to climate change, this could increase the risk of deterioration in water and waste quality. Thirdly, the emerging chemical and biological pollutants will require greater monitoring and treatment to minimise the risks to public health.
- The risks to public health associated with the poor provision of waste water treatment services will tend to increase, particularly as a result of climate change and emerging chemical and biological pollutants. There will be an increased incidence of waterborne diseases in some regions due to climate change and the effect of the land surfaces becoming impermeable, with flooding from storm water mixed with untreated waste water. Moreover, the emerging biological and chemical pollutants will require greater effort in their treatment to minimise the risks to public health, for instance in relation to bathing waters pollution.
- The correlation between water and waste services and the incidence of water borne diseases in the drinking water supply and by direct contact or the environmental impact of waste water management and solid waste management will be the subject of increasing attention by the authorities and utilities. Although utilities nowadays regularly communicate the quality of drinking water they distribute in some regions, and sometimes other parameters regarding the quality of service to users, there is still a lack of clear correlation of these aspects with their impact on public health. The same applies to the impacts of aspects linked to the seepage of underground leachate from the disposal of solid waste and the emission of gases into the atmosphere from the combustion of solid waste. There will thus be increasing studies focusing on a correlation between water and waste services and the incidence of water borne diseases in the drinking water supply and through direct contact and the environmental impact of waste water management and solid waste management.

The following are the most important challenges that can be identified at the interface between services and the environment:

- Utilities will increasingly minimise the impact of services on the environment and on water resources due to the demands and society and also legislation, for example through the simultaneous construction of drinking water supply and waste water management systems, avoiding temporary increases in pollution when the former are implemented without the latter. There will also be a better selection and protection of places used for water abstraction, with the volumes of water extracted reduced to a minimum, particularly through reducing physical water losses in the system, and adopting an appropriate treatment and final destination for treated waste water, and greater control of discharges resulting from washing and disinfecting pipelines, and utilising best treatment practices for waste water and solid waste.
- Utilities will increasingly minimise the impact of water and waste services on energy resources, for example by choosing systems that enable greater energy conservation, with the regular carrying out of infrastructure energy audits, through the installation of micro turbines and electrical generators to recovery energy as well as the recovery of sludge for energy production.
- Utilities will increasingly minimise the impact of water services on urban and rural vegetation, for example by placing pipes some distance from existing trees, especially protected species, using technology without digging trenches as an alternative to manual digging, and planning for digging trenches based on the vegetative cycle.
- Utilities will increasingly minimise the impact of water and waste services on the atmosphere, for example by minimising greenhouse gas emissions in treatment facilities and the waste collection vehicle fleet and minimise the emissions of substances hazardous to the health of workers and individuals in treatment facilities, for example chlorine.

Finally, it should be noted that the following are the main challenges that can be identified regarding the models of governance associated with water and waste services:

- Governments, both central and local, will increasingly ensure an appropriate institutional organisation for the sectors, improving their models of governance with regard to the increasing complexity that is associated with it. They will increasingly seek to optimise the institutional models in the sectors, moving towards greater overall efficiency in the provision of water and waste services, to more easily achieve the required goals with regard to coverage, continuity, quantity, quality and tariff moderation. They will also rationalise the various sources of funding available and ensure suitable revenue generation, to maximise potential economies of scale, scope and process and thus attain sustainability for the systems and clarify the role of the various types of public or private stakeholders in the sectors and gradually promote competition.
- Governments will increasingly strengthen and improve the instruments for regulating water and waste services to encourage efficiency and protect users. Linked to that explicit regulation will increasingly be generalised and improved in order to promote greater efficiency in the provision of these services, thus ensuring that users have services with the required quality at the lowest possible price. This regulation of services will increasingly strengthen articulation with environmental regulation, public health regulation and competition regulation. The main trends in regulation will be to gradually strengthen the structural regulation of the sectors and the behavioural regulation of the service utilities, irrespective of whether or not this involves the private sector. Other trends will the increasing role of user associations and the strengthening of regulatory transparency and stakeholder participation, with increased regulatory independence.

- Governments will increasingly promote the development of national, regional and local economic activity associated with these sectors and potentially create jobs and wealth. The corporatisation of water and waste services, both public and private management, will continue to be found with demand for gains in efficiency and effectiveness. New forms of public-public and public-private partnership will be developed for example through delegation or concession management models. An increase will be seen in contracting between the different sectors' stakeholders and increasing attention will be given to the sharing of risks between the parties involved, for example the delegator or grantor. There will be a growing demand for greater transparency and social accountability of utilities.
- Governments will increasingly seek to ensure appropriate sustainability for these water and waste services particularly through the ability to generate revenue. The growing need to ensure appropriate sustainability will lead to the implementation of realistic tariffs, thus ensuring a set of principles: the defence of the interests of end-users regarding continuity, quality and cost of services; the payment ability of the users, safeguarding social concerns; transparency to users; recovery of services cost, including investment amortisation, operational costs, financing costs, return on capital and other charges, including taxes; standardising tariff structures; sustainable use of water resources and the prevention and reduction of waste production and its recovery; non-subsidisation between activities.
- Water and waste services utilities will increasingly make better use of economies of scale, for example by reconfiguring systems and promoting their integration. They will increasingly make better use of economies of scope, integrating drinking water supply with waste water management. They will also increasingly make better use of economies of process, integrating bulk management, for example water production, with retail management, for example water distribution, although there can also be a contrary trend of vertical disintegration to introduce competition.
- Governments will increasingly participate more actively in the formulation and adoption of legislation and international standardisation, due to their economic impact on water and waste services. The increasing globalisation of society and the corresponding economic and political integration tend to unify legislation and standardisation in the services, which requires a continuous effort by the utilities to adapt. This effort usually entails significant additional costs and subsequent pressure on tariffs. The debate around the added value of new legislation and standardization will tend to reinforce itself.

2.8 RIGHTS OF WATER AND WASTE SERVICES USERS

Users have rights and tend to give increasingly attention to water and waste services, particularly regarding physical and economic access to services, their quality, the quality of drinking water, information about services, complaints about services and their participation in decisions (Figure 2.5).

Each of these rights is analysed in more detail below and the instruments that must exist in order for these to be assured:

Right to physical access to the services

The rights of users for physical access to the services depends meaningfully on the existence of a suitable national strategy to ensure access by the population, as well as legislation to ensure the right to be connected to the system, the right to have available those services upon request and the right to continuity of services, which can only be interrupted for exceptional reasons.

Figure 2.5 Rights of water and waste services users.

Right to affordable access to the services

The rights of users for affordable access to the services through the setting of tariffs in keeping with the economic capacity of populations, monitored through macroeconomic accessibility indicators, for example with the existence of a volumetric tariff with progressive blocks, the first being low cost, without autonomous payment of the service connections but inclusion of their costs in the tariffs, through a social tariff for economically disadvantaged users, with the family tariff and with it being forbidden to require a guarantee from domestic users before the service provision.

Right to quality of services

The right of users to quality of services involves the annual assessment of the quality of service for each utility, by comparing that with the previously established quality of service reference values, by comparing the quality of service between utilities (benchmarking), by assessing the evolution of quality of service over time, by periodic inspection of the utilities, by monitoring and eventually penalising any occurrences of non-compliance and by the availability of information.

Right to water quality

The right of users to quality drinking water involves ongoing assessment of the quality of water supplied by the utility, by comparing that water quality with previously defined reference values, by comparing the quality of service between utilities (benchmarking), by assessing the evolution of quality of service over time, by periodic inspection of the utilities and the analytical laboratories, by monitoring and penalising any occurrences of non-compliance and by the availability of information.

Right to information

The right of users to information about services involves the provision of credible and easily interpreted information from the utility of the services, tariffs, quality of service, water quality, complaints and also through the provision of the service regulations and contracts and frequently asked questions and responses.

Right to complain

The rights of users to complain about the services entails the right to complain through appropriate instruments and have a complaint answered in a proper and timely manner, through the right to independent intervention by the regulator in this matter and the right to permanent consultation regarding the state of their complaint.

Right to participation

The right of users to participate in main decisions about the water and waste services involves the presence of their representatives in key decision-making processes for the sectors, for example by participating in the advisory body of the regulatory authority and participate in the public consultation processes launched by governments, regulators and other public administration.

In terms of civil society, users are normally represented by consumer protection associations, at a national, regional or local level, which must have a legal personality, are non-profit and have the main purpose of protecting the rights and interests of consumers in general and their associates.

Consumer associations shall enjoy the following rights, among others: represent consumers in consultation processes and public hearings held in the course of making decisions likely to affect their rights and interests; consult specific documents and other elements in the central, regional or local administration services which contain data on the characteristics of the services and to disclose any information necessary for the protection of consumer interests; be informed about the pricing of the services, when so requested; participate in the economic regulation processes for services provision and request clarification regarding the tariffs practised and the quality of the services in order to be able to express an opinion concerning them; request official laboratories to carry out analyses on the characteristics of the water and make public the corresponding results; report and constitute themselves as witnesses in a criminal process and monitor the administrative process where so required.

2.9 SUMMARY

In this chapter the water and waste services were described, considering their relationship with public service obligations, the characteristics of these services, sectors' stakeholders, the service systems, the environment and water resources, the challenges for services and user rights.

The next chapter will consider the respective public policies, as tools for their implementation for the benefit of society.

Chapter 3

Public policies for water and waste services

3.1 INTRODUCTORY NOTE

Drinking water supply, waste water management and solid waste management have taken on increasing importance in the global context, and are essential public services for the social and economic development of any country. Since their characteristics are partially distinct, the water services can be defined covering drinking water supply and waste water management, and waste services, covering solid waste management.

They have major implications for the environment and public health, and it can be verified that the healthiest societies with the longest life expectancy are those countries and regions with high level of care regarding these services. On the other hand, the highest death and mortality rates occur mainly in countries and regions with an insufficient level of care regarding these services. Hence, these sectors have been clearly earmarked as a priority by policy makers around the world and this had led to public opinion being increasingly attentive in this regard.

It should thus be the objective of all countries to promote the development of these services in order to improve the quality of life of local populations. These are however services with high investment and operational costs. It is essential to ensure that such development is sustainable, being carried out in a globally consistent manner, with proper coordination of all important aspects. This implies the existence of suitable public policies for the sectors that make use of a whole range of instruments of various kinds, globally competing for the same purpose and guaranteeing an optimisation of results in relation to the resources available. This will, by this means, avoid huge investments being made with no guarantee of obtaining the expected benefits for society.

It is thus a responsibility of governments to create the necessary conditions for the gradual generalisation of access of the entire population to these water and waste services, and conditions that must necessarily define appropriate public policies.

3.2 INTERNATIONAL FRAMEWORK

Internationally there have been many policy initiatives to encourage governments to give greater attention to water and waste services.

Of note are the Millennium Development Goals adopted by the United Nations in 2000, which set targets for water services in terms of population coverage. This document set the goal of countries halving their population without access to safe drinking water and sanitation by 2015, which has not yet occurred in many situations. New Millennium Development Goals are now under discussion.

Additionally, in 2010 the General Assembly of the United Nations declared access to safe drinking water and sanitation as essential human rights for the full enjoyment of life and for all other human rights, thus reinforcing the importance and concern that increasingly fall on these sectors.

These mean the right of all citizens to have access to water services which are appropriate and safe, essential to public health and the protection of the environment, through traditional collective systems, simplified collective systems or individual systems. The services, as human rights, must be physically accessible, appropriately sized, hygienically safe, affordable and culturally acceptable. Citizen participation in decisions, access to services without discrimination, monitoring of the situation and its regular reporting should be ensured.

These rights also mean that member countries of the United Nations have the obligation to carry out the necessary measures to achieve them. The pursuit of these rights by member states, through their governments, means the obligation to respect, not threatening nor limiting access, the obligation to protect, that is, prevent threats or limitations by third parties, including providers of water services, and the obligation to fulfil these rights by supporting citizens in their access, promoting basic hygiene care through education and actually providing access to the most vulnerable citizens.

It is important to clarify that this United Nations resolution does not mean however that the member states have to provide these services immediately to their entire population, because this is generally not feasible. They do not have to provide these services directly themselves, because may do so through different stakeholders. They do not have to provide them for free, since they can and should be associated with a tariff.

These rights mean, to sum up, that the member states should define and implement appropriate, coherent and integrated public policies for water services, and also assume a major commitment regarding their implementation.

The Millennium Development Goals and the United Nations resolution about the access to safe drinking water and sanitation as essential human rights should not be seen as an end in themselves, but as opportunities for additional progress regarding universal access. Their implementation requires long-term planning, involves progressive realisation and will be highly financially and logistically demanding.

Even without being considered a human right, the services of solid waste management can benefit, through their similarity, from this new framework for the water services.

3.3 PUBLIC POLICY COMPONENTS

3.3.1 Overview

In general terms, any public policy concerning access to drinking water supply and waste water management, and similarity concerning access to solid waste management, should form a global and integrated, that is to say holistic, approach and include several components, which are described below:

- Adoption of strategic plans for the sectors
- Definition of the legislative framework
- Definition of the institutional framework
- Definition of the governance of the services
- Definition of the access targets and the quality of service goals
- Definition of the tariff policy
- Provision and management of the financial resources

- Construction of the infrastructure
- Improving the structural and operational efficiency
- Human resources capacity building
- Promotion of research and development
- Development of the economic activity
- Introduction of competition
- Protection, awareness and involvement of users
- Provision of information

The successful implementation of public policies for water and waste services depends on the ability to manage the implementation of all these components at relatively the same time, ensuring an effective global and integrated approach (Figure 3.1).

It is important to note that the implementation of one or a subset of these components is generally not sufficient to ensure the sustainability of the sectors, as this does not achieve the results expected in a long-lasting manner. For example, if it is possible in any region to make available and manage the financial resources to enable the construction of infrastructure, using funds from cooperation and development assistance, but without being provided with the other components, namely a good legislative framework, a suitable institutional framework, good models of governance of services, an effective tariff policy and human resources capacity building, then the odds of success are very small and the investments made will not have the expected return.

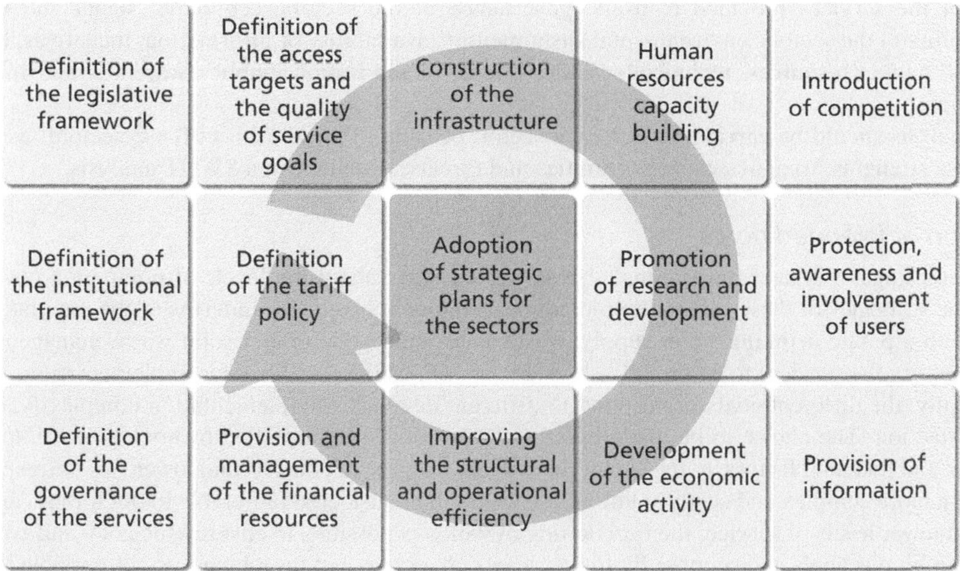

Figure 3.1 Public policy components.

Moreover, it is necessary to take into account that the implementation of public policies for water and waste services should be progressive, due to their great complexity and the costs involved, focusing on the priorities of the country and giving special attention to the neediest users.

These various components are described in more detail below.

3.3.2 Adoption of strategic plans for the sectors

The political, social and economic importance of these sectors requires that there are appropriate strategies, preferably expressed within a strategic plan for the sectors for drinking water supply and waste water management, and other strategic plan for the sector for the solid waste management, nationwide in scope and in the medium term, corresponding to the government's vision for these sectors and for society.

Nationwide strategic plans for the sectors, for rapid but smooth development of the sectors, involve stages characterising and diagnosing the current situation, defining intended goals, evaluating the corresponding investment needs, identifying the measures needed, implementing activities and specifying the monitoring instruments, which are detailed below:

Characterisation and diagnosis of the situation

Initially, it is necessary to make a general description of the situation of the country with interest to the sectors. Without being exhaustive, this description should include aspects such as demographics and population spread, assessing public health levels, the level of coverage of these services, an inventory of existing infrastructure elements, quality of the services, water needs and availability, production of waste water and solid waste, technical and financial capacity of the utilities and the financial capacity of users.

This description should be complemented by a diagnosis of the specific situation for the sectors, addressing aspects such as the legal framework, institutional framework, physical accessibility to services, quality of the services provided to users, governance of the sectors, economic, social and financial sustainability of the sectors, environmental sustainability, availability of information, incentives, financial resources, human resources, technical resources, level of innovation, public awareness and impact on employment.

An analysis should be carried out for each sector, bearing in mind each of these sectoral aspects, to identify its strengths, weaknesses, opportunities and threats, which mean a SWOT analysis.

Definition of intended goals

In a second stage, it is necessary to make strategic decisions about the levels of population coverage to reach. The strategies of these sectors should advocate national objectives in terms of the population to be served with a public drinking water supply, waste water management and solid waste management, the level of quality the services to be provided to users, as well as the deadlines for implementation.

Naturally, the different goals correspond to different degrees of implementation complexity, cost and user satisfaction. The choice to be made by a particular goal is conditioned by sociocultural, technical, economic and financial factors, to the extent that a higher degree of comfort and lower risk corresponds, as a rule, to a more complex and costly solution. It is generally considered preferable to opt for less ambitious goals and lower levels of service, the functioning of which is possible, to ensure efficiency and continuity, to more ambitious goals and supposedly higher levels of service, but for which it is not possible to ensure sustainable functioning.

Evaluation of the corresponding investment needs

Once the actual situation and the intended objectives are known, it is then possible to define the needs, particularly in terms of investment, operation and rehabilitation, and consequently estimate the sums involved, which in general are very high and constitute one of the limiting factors for development in these sectors.

Identification of the measures needed

In a fourth stage, a set of measures should be identified and developed at the institutional, legislative, economic, financial, and technical level, along with the innovation and public awareness that are considered necessary to be able to sustainably develop the strategy for the sectors. These measures should include the key factors of change, describe measures that are deemed necessary to correct the problems identified in the diagnosis and recommend the best way to implement them.

Implementation of the strategy

Finally there is the implementation stage of the strategy, through the measures identified in order for the goals defined in the second stage to be achieved. It should be noted that the implementation of these measures has always to bear in mind the existence of a dependence of these sectors regarding other external and broader sectors such as water resources, the environment, public health and land use, so success also depends largely on these other sectors.

Definition of the monitoring instruments

The strategy should include monitoring instruments that enable the regular monitoring of their implementation, identifying any problems and the need to introduce any possible changes.

3.3.3 Definition of the legislative framework

States should set up an appropriate legislative framework to cover the legal regime for water and waste services and their regulation, as well as tariffs, quality of service, quality of drinking water and also the relevant technical aspects.

Additionally, it is important to make available suitable legislation on water resource management, waste management, environmental management, consumer protection and competition.

In each case there must be a periodical analysis, balance and evaluation of the existing legislation, resulting in the identification of issues to be reviewed, particularly gaps, inconsistencies and outdated aspects, and the complementary aspects requiring new legislation.

3.3.4 Definition of the institutional framework

States should establish an appropriate institutional framework, with a clear assignment of responsibilities for the public entities involved, especially the regulatory authority for water and waste services and the environmental, water resources, waste, health, consumer protection and competition authorities, without prejudice to the possibility of other institutional arrangements being set up to fulfil these tasks.

This definition is absolutely essential for a good performance by the sectors, since it enables the responsibilities of stakeholders to be specified, along with clear rules of action and the articulation between the close and complementary sectors mentioned above, without undesirable overlaps or gaps.

3.3.5 Definition of the governance models for the services

States shall establish governance models that can be used to provide the water and waste services, for example, by direct management, delegated management and concessions.

Policy makers should also explore the general model to be adopted for the sectors, particularly in terms of its openness to the private sector, which can range from maintaining these services under public responsibility and management, maintaining the systems under public responsibility but handing over their management to private entities, and privatising the services, which means handing over responsibility and management to private entities. The public-private discussion is recurrent but often sterile, and detracts from the public policy matters that are important in the achievement of effective, efficient and sustainable water and waste services, which are the real targets.

3.3.6 Definition of the access targets and the quality of service goals

States should set realistic access goals, namely the population that must have available public water and waste services, and quality of service objectives, with subsequent regular monitoring. It is necessary to ensure a sense of realism in the setting of the goals, and to oversee the process of turning them into deliverables.

The way in which utilities have implemented this goal has evolved over time. The concept of good quality service has historically tended to pass through three stages, namely the quantity stage, in which the main task was to satisfy the basic quantitative needs of the population, the quality stage, where the water quality objectives were joined to the previous stage, and the excellence stage, which seeks to add the strand of sustainable development in social, economic and environmental terms to the quality stage. Access goals and appropriate quality of service objectives have to be found which are suitable for the specific situation of the country.

3.3.7 Definition of the tariff and tax policy

States should set out a tariff policy for public water and waste services that promotes a gradual trend towards cost recovery, consistent with the economic capacity of the population.

It is necessary to gradually implement a tariff policy that is as realistic as possible, within a framework of the principle of the user-payer, without prejudice to implementing a system of brackets involving social considerations. It is therefore necessary to avoid average tariffs below actual values or even the absence of tariffs, which prevents the generation of revenue for the utility and therefore creates difficulties and the inability to cover the depreciation costs of investments and operation.

At the same time the introduction of tax instruments may be suitable, which may encourage desirable behaviour, for example in the rational use of water as a raw material or as final usage, through a water resources usage levy, or as final destination suitable for solid waste, through a waste destination levy.

3.3.8 Provision and management of the financial resources

States should ensure the provision of the substantial financial resources necessary for the sectors, or via cooperation and development support funds, and empower itself for the efficient management of these.

When sources of funding are available for water and waste services, which entities can apply for these and the eligible projects and priority projects should clearly be defined, in order to optimise investment in the sectors.

As for eligible entities, from the outset projects submitted by any utility of water and waste services should be considered for funding regardless of their management model, which may correspond to direct, delegation or concession management. It is important that the eligibility criteria are not discriminatory on the basis of the management model adopted for the provision of services and conditions are not imposed as limitations from the corporate standpoint or capital ownership of the managing bodies. The use of eligibility criteria as a way to constrain the choices of the management model is therefore not recommended, as these are the responsibility of the entities holding these services, with the focus being on an analysis of the bids based primarily on the merits of the proposed technical solutions.

As for the eligible projects, in general the following criteria for access to financing from cooperation and development support funding should be considered, for example:

- The project must meet the national goals defined for the sectors, ensuring full compliance with the strategic and operational objectives laid down for the coming years. It should in particular be compatible with the objectives of water resources planning for the respective watershed or solid waste planning, and should be contextualised in the programmes of measures aimed at improvement and environmental protection.
- The project must fit within general plans for public drinking water supply, waste water management and solid waste management. These plans must include the detailing of existing infrastructure, their operational status, envisaged extensions and enlargements, the respective investment and operational costs and an implementation schedule.
- The project should demonstrate investment optimisation from the public interest perspective, proposing appropriate levels of integration with existing technical solutions or their construction to enable the generation of economies of scale in investment and operation, involving territorial integration, economies of scope, with integration of drinking water supply with waste water management and economies of process, with the integration of the production process. This quest for integrated management should be naturally compatible with the adoption of solutions which are not physically integrated, whenever these are justified, particularly in rural areas.
- The project must be able to demonstrate financial and economic sustainability, ensuring the necessary revenue to cover all investment and operating costs.
- The project should demonstrate an appropriate management model that allows for technical, economic and environmentally qualified operation and offers guarantees of compliance with legal and contractual obligations.

Finally, the projects that should be prioritised for funding from cooperation and development support funds are those that:

- Demonstrate they will be able to maximise the benefit of existing investments, independent of whether these systems are under the management of a single entity or separate entities.
- Demonstrate they will be able to achieve an important improvement in the quality of service provided to users, particularly an important improvement in the quality of drinking water, especially when public health aspects are involved.
- Demonstrate it will enable an important improvement in environmental terms, particularly with the more rational use of water and protection against pollution, particularly in problematic areas.
- Demonstrate that funding through grants is essential to enable an average tariff acceptable for the level of economic and social development of the population, thus avoiding an excessive tariff that would result from the strict application of economic and financial sustainability criteria for investments only recoverable by means of the tariff.
- Are presented by utilities that show an ongoing and consistent effort to comply with legislation, including environmental and public health measures, and compliance with any existing contractual arrangements.

Regulations regarding accessing the financing funds should be imposed as indispensable conditions for fund allocation. For example, aspects such as economic and technical sustainability, the human and organisational resources of the utility, the minimum physical size of the systems to be financed, compliance with existing legislation in the field, the existence of a master plan, the existence and regular updating of programmes for operation, maintenance, security and training, the existence of suitable inspection of the works, the use of qualified entities and the adoption of a self-assessment system for the quality of service provided to users.

3.3.9 Construction of the infrastructure

States should promote the construction of the infrastructure for the provision of public water and waste services, using appropriate technology, naturally implying substantial costs, both in terms of initial investments and at the operational level.

The financial sustainability of the services is crucial, as the first way to ensure quality services to the users. To this end, the tariffs should match the need for self-financing. Financing in the form of grants may, however, be exceptionally allowed, in the case of the so-called first infrastructure, which is socially justified through their high cost not being passed on to users, with the costs being distributed among taxpayers at the general level of the region, the country or third countries.

As a consequence, appropriate funding mechanisms, as mentioned above, are naturally indispensable for good performance by the sectors, using available sources, consistent with the objectives which have been established.

3.3.10 Improving the structural and operational efficiency

States should promote the improvement of the structural efficiency of the public water and waste services and the efficiency of the operation by the utility.

In improving the structural efficiency of the sectors, an optimised territorial organisation should be specified for the management of these services, making use of economies of scale, normally at the regional level. The utilities should therefore promote, as much as possible, the physical integration and aggregation of the drinking water supply, waste water management and solid waste management systems at a technically and economically appropriate scale, promoting where necessary the implementation of joint solutions with similar entities in order to cover a territory that does not have to be necessarily confined by administrative boundaries. In fact, although physical systems integration is profoundly influenced by local conditions, there are clear benefits resulting from the physical aggregation of systems, not only in economic terms but also in terms of the quality of the services provided.

In improving the efficiency of the operation of the utilities, they should seek to adopt, given the existing legislation, the most advisable organisation-type, particularly at the staffing level, the functional content, information circuits, administrative procedures, planning, budget, control and measures which will tend to ensure quality assurance.

3.3.11 Human resource capacity building

States should promote the human resource capacity building in terms of number and skills. Indeed, the existence of human resources with suitable technical training to carry out their duties linked to public water and waste services is an essential factor in ensuring overall quality.

In order to achieve this there should be a strengthening of technological and traditional technical courses (university and polytechnic courses) to overcome any shortages of personnel with relevant academic qualifications to carry out the necessary roles in the sectors.

There should also be training and updating courses for specific human resources of the sectors, based on a preliminary needs analysis and the type of training. These programmes presuppose the existence of a network of trainers and training activities should be offered for managers and technicians at various levels, internships at home and abroad and even technical exchanges between utilities and research bodies.

There should be measures to encourage the edition of technical publications and various other educational material of a practical character in the native language, covering all sector areas, aimed at the different professional levels involved and which will be used in particular to support the aforementioned training activities.

3.3.12 Promotion of research and development

States should promote research and development in areas associated with public water and waste services, creating endogenous knowledge and thus ensuring increased national autonomy.

A research and development program for the sectors should be defined, covering pre-normative applied research projects, for the development and strengthening of the technological infrastructure on which they can base these projects.

There should be a programme for technology transfer in these sectors, through short internships for technical administrators and operators in research and educational institutions, and conversely for researchers in administrative institutions and utilities so that they can build human capacity using the new approaches and innovative tools available.

3.3.13 Development of the economic activity

States should promote the development of the economic activity related with these sectors. Effectively, when implementing strategies to develop water and waste services, this creates exceptional conditions to promote the development of national knowledge and hence strengthen the capacity of the economic activity in the market. Calling on society to collaborate, particular the business sector, it is possible to generate new activities by creating jobs and wealth.

3.3.14 Introduction of competition

Given the monopolistic characteristics of the sectors, competition in public water and waste services should be promoted as much as possible, in so far as it is a motivator for innovation and technical progress and, therefore, for increasing the efficiency and quality of its provision.

In the case of natural monopolies, and where there is therefore no competition in the market, virtual competition should be promoted, for example through benchmarking between utilities. It is an effective substitute for direct competition in a monopoly situation.

In the case of private involvement, competition in the market should be promoted, for example through tender procedures for the allocation of delegations, concessions and the provision of services.

It is important to minimise the monopolistic characteristics of the sectors and the risk of dominant position abuse and other anti-competitive practices which are contrary to the interests of users.

3.3.15 Protection, awareness and involvement of the users

States should promote the development of appropriate tools for the protection of users, especially the most economically disadvantaged, for awareness increase and for improvement of public participation in decision-making with regard to water and waste services. There should be environmental education activities with the provision of educational materials.

3.3.16 Provision of information

All these stakeholders, to a greater or lesser extent, need to have reliable information about public water and waste services and their development, both to support the definition of public policies and business strategies and to evaluate the service that is actually provided to society, so as to be able to convey an overview of the sectors in a reliable and regularly updated manner.

Suitable information regarding the sectors, understood as a set of data on service levels, the characterization of the utilities and users, consumption, service levels provided, etc. constitutes an essential tool for defining development strategies and for supporting various sector activities. It is furthermore essential that there is proper monitoring of the sectors growth by the public administration. Any omissions or errors of information can have a severely negative impact on decision-making at various levels, particularly regarding investment planning.

Information is a necessity at various levels, and in particular the following two. The first is at an essentially national or regional level, most useful for defining policies and development strategies, and a second essentially at the level of the utility, most useful for the operation of the systems.

Regarding the first group, there has to be a suitable statistical system with a clear allocation of responsibilities to a single public entity, in principle the regulatory authority, in relation to the execution and coordination of this function, with the task of periodically collecting, validating, filing and processing information about the situation of the sectors, following a sufficiently standardised format, which must be sufficiently complete and easy to interpret.

For the second level it is necessary that the utilities regularly record and use this information and make it available in a timely manner, which means in many situations a cultural change.

3.4 ROLE OF REGULATION IN PUBLIC POLICIES

Regulation should be seen as a component of public policies on water and waste, one out of various, but which has a very important role given the fact that it promotes or controls most of the remaining components. It can be seen as the procedure of interpreting and implementing laws, policies and regulations, to achieve the global objectives for the water and waste services.

In most situations it is intended that the State is the promoter of the effective and efficient delivery of essential public services, at an appropriate level of quality, at social affordable prices and an acceptable level of risk for users, while simultaneously ensuring the economic, social and environment sustainability of the providers of these services. The aim is to find the optimal balance between these goals, particularly to thereby ensure transparency towards users, regardless of whether the governance model is public or private.

When discussing the roles that may be assigned to the State, whether in a more minimalist view, focusing on essential functions, or in a more maximalist vision, including direct intervention in the economy, what emerges as a consensual aspect is the fact that regulation is one of the main functions of the State, albeit that the proposals for the model and level of intensity to be used in such regulation often diverge.

However, regulation is not the solution to all problems but only a relevant instrument of public policy, and is more effective according to the level of consistency of that policy.

3.5 SUMMARY

In this chapter public policies have been described for water and waste services, covering the international framework, the various components of a public policy and the role of regulation in that public policy.

The following chapter will consider some of the good practices to adopt in the creation or reorganisation of a regulator for public drinking water supply, waste water management and solid waste management services to populations, essential for the implementation of the regulation in a beneficial manner to society.

Chapter 4

Setting up a regulatory authority

4.1 INTRODUCTORY NOTE

Following the integrated regulatory approach (model RITA-ERSAR) which has been advocated, the present chapter presents some good practices to adopt in the creation or in the reorganisation of a regulatory authority for public water supply services, waste water and solid waste management.

4.2 NEED FOR REGULATION OF SERVICES

As has already been mentioned, the public water and waste services sectors are configured from the point of view of a market structure which is in a situation of a natural or legal monopoly. This situation is especially evident in the cases of the drinking water supply and waste water management services. These are typical network industries in which the activities are carried as a natural monopoly of a local or regional kind, probably the last of the great monopolies in public services.

Other network industries are electricity, natural gas, telecommunications and railway transportation services, and, similarly, air and sea transport and the postal services.

Given that the monopoly is a fault in the market, in the sense of it not being competitive, regulation arises as a way of reducing the forecast inefficiencies resulting from this fault in the market and the potential reduction in social well-being as regards the decreased safeguarding of the public interest.

When there is a public service which constitutes a natural monopoly, the State, in addition to being the service holder, may ensure that the provision of that service through central, regional or local utilities satisfies the public interest and the needs of the users, without taking advantage of its monopolist condition. The State can try to avoid a reduced quality service at a higher price than that which would result from a competitive market. This option implies, however, that the State is interested and actually capable of effectively and efficiently providing that service, with the public interest in mind.

But this does not always happen due to the State's difficulty, at central, regional and local levels, of focusing on and assuring capacity for the effective and efficient provision of water and waste services, or due to ideological and strategic options taken by the State, for example deciding that such services should be managed by third parties with private capital. Even when the State delegates or grants these services to public companies, experience shows that other priorities such as the creation of shareholder value frequently override expected concerns regarding public service obligations.

When the State cannot directly provide an effective and efficient service, due to difficulties or through choice, monopoly markets require a form of regulation that goes beyond the lack of self-regulation, which is a characteristic of competitive markets. Without regulation there is a natural increase in the risk of the utilities prevailing to the detriment of users, since they will inevitably seek to profit from their dominant position or market power, resulting in lower quality services and higher prices for users.

Regulation is therefore a mechanism which seeks to reproduce, in a natural or legal monopoly market, the resulting efficiency which one tends to naturally obtain in a competitive market. It could be said that a market of virtual competition is created, introducing regulatory measures to induce utilities to act based on public interest, although naturally without risking their own viability.

Regulation thus arises as a modern instrument of State intervention in the water and waste services, which is essential for their smooth functioning and the defence of the public interest. The regulatory authority thus carries out the role of the State, specifically those it delegates in that authority. The need of autonomous regulators is reinforced by the belief the policy, regulation and provision of services should be preferably separated to give focus to the application of the required expertise, and to provide transparency.

This is the reason why we have seen the creation of a growing number of regulatory authorities throughout the world, which constitutes apparent recognition of the advantage in creating specialised bodies which exercise State powers more effectively. They are generally more effective because regulation is of a greater technical nature, which is very complex and interdisciplinary, for which it is sometimes difficult to find a solution in traditional public administration.

The alternative of self-regulation of those services by the utilities themselves does not guaranty a balanced approach and by now has failed to demonstrate in most of the situations that guaranties the provision of efficient and sustainable water and waste services.

4.3 REGULATION OBJECTIVES

As has already been mentioned, activities involving drinking water supply to populations, waste water management and solid waste management are essential public services of a structural character which are vital to the general welfare, public health and collective security of populations, their economic activities and protection of the environment These services should be guided by principles of universality of access, continuity and quality of service, as well as efficiency and equity in terms of the tariffs applied.

The main objective of the regulation of these services is thus the protection of the interests of the users through the promotion of the quality of service provided by the utilities and the guarantee of socially affordable tariffs, at an acceptable level of risk.

Care should also be taken to safeguard the economic viability and the legitimate interests of the utilities, thus ensuring in particular a suitable but not excessive return on invested capital, independently of whether their status is public or private, municipal, regional or State-owned. Equal and transparent conditions should also be ensured as regards access to the activity and in the corresponding management, as well as in terms of contractual relations. This regulation should also contribute to the implementing of public policy as defined by States and for economic development through the consolidation of support for the services of the non-regulated business sectors.

As these activities have a strong impact on environmental and natural resources, environmental protection is also an aspect that should not be neglected in the regulation of the services, without prejudice to the functions of the specific environmental authorities. This requires careful articulation with such authorities.

In summary, the rationale for the regulation of the water and waste services thus involves contributing to ensuring global sustainability, which can be categorised in three aspects:

- The social sustainability of the services, ensuring the protection of the interest of users through access to the service, at a suitable quality and reasonableness of price.
- Economic, infrastructure and human resource sustainability for the entities.
- Environmental sustainability in terms of the efficient use of environmental resources and the prevention of pollution vis-à-vis the impact of the services on water, air and on soil.

4.4 REGULATORY MISSION AND MANDATE

When setting up a regulatory authority, it is essential that the State defines with utmost clarity what is understood to be its regulatory mission, its regulatory mandate and its regulatory objectives. This clarity is essential so that the regulator is suitably legitimised and may fulfil its role in an effective manner.

The first condition of the responsibility as a regulatory authority is the precise delimitation of its regulatory mission, which should clearly mark the frontier between, upstream, legislative power and public policy options, falling under the remit of parliament and governments, and downstream, the area of intervention of the regulatory authority. The same applies in the delimitation of the jurisdiction of the various sectoral regulatory authorities among themselves and in relation to the national competition authority.

There is nothing more prejudicial to the responsibility of the regulatory authority than doubts concerning the frontiers between its area of intervention and its powers, which generate conflicts in terms of duties and which dilute its responsibilities. A clearly delimited mandate, in terms of its duties, competencies and powers is thus the first condition for responsibility.

The regulatory authority should therefore have clearly defined its mission, which may for example correspond to the regulation and supervision of the sectors involving drinking water supply, waste water management and solid waste management services. But there are good examples of multi-sectoral regulators. Clarity on responsibilities is however essential.

The regulatory authority should carry out public policies within a suitable legal framework, producing benefits which justify its own costs to the society, minimising public service costs and market distortions and promoting innovation through market incentives, in a consistent manner articulated with other public policies and with the competition.

In Portugal the bylaw of the Water and Waste Services Regulation Authority were approved by Law No. 10/2014 of the parliament, on 6 March, after intense debate with all the sectors' stakeholders, and which will be reproduced in Annex C. It represents a good balanced example of political decision-making regarding the nature, mission and the duties of the authority, its powers in terms of authority, sanctions and regulations, its organic structure and staffing, its independence and asset, budgetary and financial system.

4.5 CHARACTERISTICS OF THE REGULATORY AUTHORITY

4.5.1 Overview

This section concerns the main characteristics to be defined and good practices to be adopted by a regulatory authority within a framework of its setting up or reorganization.

4.5.2 Regulatory principles

In order to promote a more effective regulation it is essential to ensure that the regulatory authorities are concerned with principles of competence, responsibility, exemption and transparency and that these are reflected in their organisational laws, as mentioned below:

Competency

The activity of the regulatory authority should be carried out based on its own sound capacities which it should possess in order to evaluate the issues inherent to its regulation, taking specifically into account, and in an integrated manner, the technical, economic, legal, environmental, social, ethical and public health aspects which characterise these services.

Responsibility

The activity of the regulatory authority should be carried out within the scope of its mission statement to which it was entrusted and its attributes, in association with the political power and all the other stakeholders and interested parties within the sectors, in strict compliance with applicable legal legislation.

Exemption

The activity of the regulatory authority should be carried out with total impartiality with regard to the interests of the different parties involved. It should therefore be positioned equidistant from political power, from the utilities, from users and the other agents in the sectors, acting strictly in terms of public interest.

Transparency

Trust in the decisions and actions of the regulatory authority, from users, utilities and other stakeholders depends from the existence of independent and clear regulation procedures, transparency mechanisms and monitorable accountability and ongoing involvement from participants within the sectors.

Procedures should guarantee that they promote the effective and efficient carrying out of public policy objectives but also that they minimise potential conflicts of interest and eliminate the risk of the regulator being entrapped by specific interests.

The regulatory authority should provide statements of its activity in a clear and acceptable manner, specifically through an annual activities report and financial statement, information made available on its website and through means of communication and publications disseminated through available communication channels, and through its advisory board, government and parliament.

4.5.3 Regulatory independence

In addition to being guided by these principles, the regulatory authority should be independent in carrying out its functions, without prejudice to the public policies defined for the regulated sectors, in constitutional and legal terms.

This independence includes preventing external interference and also internal constrains. As regards external interference, this may occur in functional, organisational or financial terms, for both the utilities which regulate and the political powers. It is important to enable objective, impartial and coherent regulation decisions and avoid risks of partiality or bias regarding pressures from interests or the controversial and politically sensitive nature of some decisions. As far as internal constrains is concerned, these can be dealt with through administrative, human resources and budgetary management autonomy.

It is absolutely essential to establish a separation of powers between, on the one hand, the political bodies of the State, namely parliament and governments, which should of course define the legislative frameworks and strategic lines of regulatory activity and, on the other hand, the regulatory activity in itself. This activity consists of the implementation of the legal framework, within the context of the aforementioned strategic options, through the drawing up of secondary legislation, supervision and the sanctioning of infractions against the established regulatory framework. To do this the regulatory authorities should make use of any necessary legislative, administrative and sanctioning powers.

As regards the essentials of regulatory independence of the regulatory authorities, which enable them to be exempt from the guidance and control of governments, the supreme organ of the entire public administration that should be accountable to parliament, it is important to highlight the following:

Separation between politics and the economy

In the current dominant paradigm of State intervention in the economic area, it is understood that even when it is essential to maintain the public regulation of the economy, this function should not belong to governments in order to accentuate the separation between politics and the economy.

Ensure the stability and security of the regulatory framework

Independence in relation to governments ensures that the regulatory framework does not depend on either the electoral cycle or changes of government, given the immovability of the mandate of the regulators and their autonomy of action. This way fosters the confidence of stakeholders with regard to the stability of the regulatory environment, which thus remains protected from unpredictable changes.

Favouring professionalism and political neutrality

Regulation is essentially a technical question, and should be as distanced as much as possible from political debate. Such understanding favours the recruitment of professional specialists instead of political cadres, thus providing guarantees of greater neutrality and objectivity for regulatory activity.

Separate the entrepreneurial State from the regulatory State

Even in political frameworks involving liberalisation, that is, the opening up to private companies of sectors formally featuring public monopolies, and also privatisation, that is, the divestiture of public companies, the State continues to have considerable direct intervention in several of these sectors. This intervention occurs while sometimes still in a dominant position, through State-owned companies or State holdings in joint ventures, or even through privileged shareholdings in privatised companies. In these cases, it is important to carefully separate the State functions of the utility from that of the regulator, so as to ensure the impartiality and equality of regulation with regard to public and private utilities.

Protection against regulatory bias

One of the greatest risks of regulation consists in the possibility that the regulator is biased by those that it regulates, in such a way as to become a form of self-regulation through means of an interposed regulatory authority. As such, the setting up of independent regulators, removed from the actual constrictions of party conflict and the electoral cycle, provides better conditions to resist any pressures from those regulated which can happen in relation to direct or indirect administration services dependent on governments, in the sense of bending regulation in their favour.

Facilitate self-financing

The organic and functional independence of the regulatory authority facilitates its financial independence, both as regards its sources of funding, via endogenous funding through levies on its own services, and as regards budgetary management, which normally includes reinforced autonomy in comparison with those so-called autonomous funds and services. This autonomy strengthens in turn its independence in terms of the relationship with governments as well as in relation to those regulated.

When talking of independent regulation what is usually meant is independence in relation to governments. However, regulatory authorities, whether they are independent or not in this sense, should always be independent with regard to regulated interests. This aspect is as much or even more important than independence in relation to governments.

The mechanisms aimed at ensuring this other aspect of independence and minimising the risks of regulatory bias are mainly the following:

Organisational independence

The designation of the management body for the regulatory authority should be made from among people of recognised repute, independence and technical and professional competence. It should be done, for example, through the collegial decision of governments following a proposal from the minister responsible, but obligatorily proceeded by a statement from parliament, after analysis of the curricula and justification of the respective choices. This ensures greater public scrutiny and more democratic legitimacy. This appointment may alternatively be made by parliament itself or by the actual head of State.

The mandate and statute for the members of the management body must be relatively long to ensure regulatory stability, for example for a period of between five to ten years without the possibility of renovation but with a guarantee of immovability.

The members of the management body may only be relieved of office before their appointed period in exceptional cases envisaged under law. The collective termination of the mandates of the members of the management body may only happen following a suitably justified government resolution, based on an investigation carried out by an independent body and with parliament having been consulted, and due to reasons of grave irregularities in the functioning of the regulatory authority. Individual termination of a mandate may only happen in cases of resignation, death or permanent disability to carry out the role, supervening incompatibility, conviction for an intentional crime which prejudices the reputation by carrying on in the role, clear and justified unfitness to carry on in the role or serious misconduct.

The loss of organisational independence of the regulatory authority, in whole or in part of the members of the management body, may be the result of bias through the appointment by governments of persons who, regardless of any technical and professional skills, do not possess sufficient repute or independence.

The main guarantee of the independence of the regulatory authority lies in the personality of the members of the management body. The more personal authority and prestige they have in the sectors being regulated, the greater their independence will be in relation to external pressures. Therefore, the law should suitably define the personal requisites and the way in which the members of the management body will be appointed. The dissemination of the curriculum vitae of those nominated and their prior presentation to a competent parliamentary commission may help to reinforce an image of independence. Rather than legislating on independence, it is necessary to practise it.

Functional independence

The regulatory authority must be independent when carrying out its duties, not receiving specific instructions from governments, without prejudice of course to its duty under law to the strategic guidelines for the sectors, in constitutional and legal terms, and duty to the minister responsible in certain specific acts, pursuant to that envisaged in law and in its bylaw.

The loss of functional independence of the regulatory authority may stem from various origins and take on various forms, since the risk of bias may come from governments, from State-owned, municipal and private-public utilities and even from users, through their defence associations.

With regard to governments, to the extent that the regulatory authority carries out the functions of the State delegated in it, there may be a potential tendency to control or influence certain regulatory decisions to benefit political interests. Should this happen, it means that governments do not actually want independent regulation, at least in certain aspects, but rather wish to keep an image of independence for the sectors.

State-owned utilities may be potentially tempted to control or influence certain regulatory decisions for the benefit of their own interests, encouraged by their position which is often a major or even predominant one within the sectors. There is also the fact that as they belong to the State, they will consequently be close to governments, which will tend to increase this temptation.

As regards municipal public utilities, there may also be a potential temptation to control or influence certain regulatory decisions in benefit of their own interests, especially when they are linked together in associative structures, encouraged by the great political weight which they in general have and by their argument of direct democratic legitimacy, which is naturally legitimate but which does not do away with the need for regulation.

There may also be the potential temptation among private utilities to control or influence certain regulatory decisions to benefit their own interests, especially manipulating market operating rules.

The potential temptation among consumers may be to control or influence certain regulatory decisions to benefit their own interests, however in a more sporadic manner due to their greater dispersal, even in associative terms, and their lower technical and even intervention capacity.

The consequences of any possible loss of independence of the regulatory authority, even if temporary, is very serious in terms of the confidence which other stakeholders place in it and this inevitably penalises its regulatory legitimacy and credibility. There is neither legitimacy nor credibility without effective independence for the regulatory authority, and consequently there is no regulation effectiveness.

Financial independence

The regulatory authority should only have its own revenues coming from levies applied to the utilities regarding its regulation activity, and it should therefore not be dependent on the budget of States and consequently of governments.

These levels of independence should be complemented with autonomy in administrative, human resources and budgetary management terms, specifically in terms of its ability to take decisions regarding its human resources and the purchase of goods and services. This is important for the effectiveness and efficiency of its activity as a regulatory authority, as systematised in Figure 4.1.

The loss of budgetary and financial independence of the regulatory authority may result in its bias towards government in the case of its own revenues, through the imposition of rules which condition their use in an unacceptable manner, and, in the case of revenue from the budget of States, through the reduction of that income to below an acceptable level.

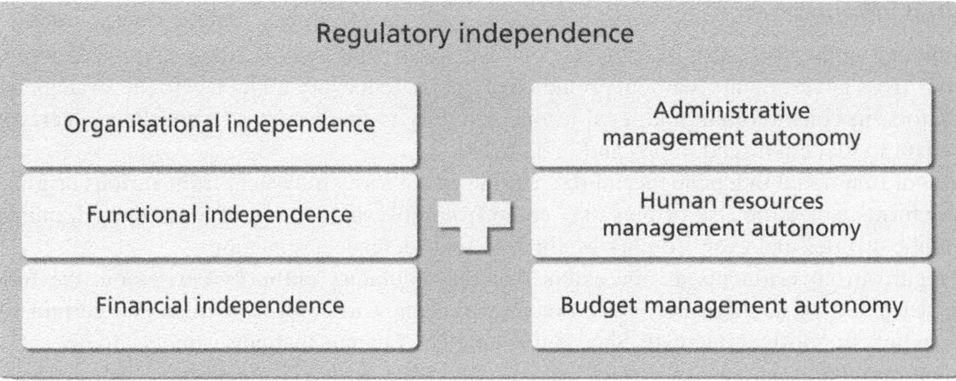

Figure 4.1 Regulatory independence levels.

The need for effective organisational, functional and financial independence, both with regard to governments and with regard to utilities and users, advocates intense scrutiny by the competent authorities as an instrument of rebalancing and legitimation. This scrutiny, particularly by parliament, will occur through accountability tools, which will be detailed later on.

4.5.4 Regulatory duties

The duties of the regulatory authority should for example be the regulation and supervision of services for public drinking water supply, waste water management and solid waste management, promoting an increase in the efficiency and effectiveness of their delivery.

These duties should be implemented through structural regulation of the sectors and the behavioural regulation of the utilities.

In terms of the structural regulation of the sectors, the utility should:

- Contribute to the organisation of the sectors;
- Contribute to the legislation of the sectors;
- Contribute to the information of the sectors;
- Contribute to the capacity building of the sectors.

At the level of the behavioural regulation of the utilities providing water and waste services, the regulatory authority should:

- Carry out legal and contractual regulation;
- Carry out economic regulation;
- Carry out quality of service regulation;
- Carry out drinking water quality regulation;
- Carry out user interface regulation.

4.5.5 Regulatory powers

In carrying out its duties, the regulatory authority assumes the powers which were granted to it by the States, generally the following:

Powers of authority

The regulatory authority should be able to exercise the powers of authority necessary to carry out its duties mainly through its powers of supervision, which is considered the primary and permanent activity of regulation.

When necessary, supervision may include the carrying out of inspection activities and auditing activities, which are articulated and integrated procedures.

Inspections are activities of a secondary and intermittent nature, involving sampling, carried out within the scope of the inspection powers of the regulatory authority. They serve the purpose of verifying the level of compliance of contracts concerning the management of services, legal and legislative standards referring to the scope of intervention of the regulatory authority, as well as other regulatory instruments defined by this entity. These are based on an extensive analysis of records and support documents relating to the activity of the inspected utility, within the period being covered, without prejudice to the use of other methods, such as interviews and questionnaires.

For example, a budget inspection may include the verification of an approved budget and the confirmation of whether it was carried out in conformity with forecasts, authorisations and regulations.

Auditing actions are activities of a secondary and intermittent nature, involving sampling, carried out within the scope of the inspection powers of the regulatory authority. Their purpose is to issue a suitably considered opinion, which may include considerations of a qualitative nature. They are based on the methodological examination of the situation, activity, function, programme or system of a particular body, in which the material validity of the elements to be controlled is ensured, verifying the compliance of the processing of the facts with the rules, standards and internal control system procedures. The aim is to issue a considered opinion through a report on global compliance of the audited object with specific objectives, principles, rules and standards. It uses specific commonly accepted techniques, in particular, sampling.

They may be carried out by external bodies contracted by the regulatory authority, which base their performance on high standards of quality and independence criteria, or, in some situations, by the utilities in accordance with criteria validated by the regulatory authority.

In this latter case, the audits carried out by the utilities are undertaken by external independent auditors of well-known repute. The content of the audits and the criteria for selection of the bodies to carry out the audits are approved by the regulatory authority, following the proposal presented by the utilities. The audit reports are sent to the regulatory authority.

There are various forms, including audits of an internal, external, accounting/financial, compliance, operational, result-based and performance-based nature.

For example, a regularity/legality audit may include the verification of the accounts submitted by the entities obliged to do so, which includes the examination and analysis of accounting records and the expression of an opinion regarding the same, the auditing of financial operations and systems, as well as the auditing of compliance with legal provisions and applicable regulations and the internal control systems.

The human resources of the regulatory authority, as well as the externally credited collaborators, should be equivalent to agents of authority and they should be entitled to, specifically, freely access all facilities, infrastructure and equipment of the utilities, and be able to remain there for any necessary length of time, and use facilities made available by the utility. Such facilities should be suitable for the carrying out their duties in decent and effective conditions. They should be able to request and copy documents as well as collect samples, equipment and materials for the carrying out of analyses and tests, consultation, support and the attaching to reports, cases or minutes and, furthermore, carry out an examination of any relevant element. They should be able to rule on the suspension or termination of activities and the closure of facilities. They should also be able to request the cooperation of the competent authorities, especially the police and administrative authorities, when necessary for implementing their duties.

The bodies subject to inspections and audits should provide all documental and verbal information which is requested of them by the regulatory authority within the specified period, and carry out any other necessary formality concerning proving, researching, examining or testing so as to prove compliance with the legal provisions.

The preliminary report should always be submitted to the utility and the concession holder so that they can issue a response and the final report should be sent to the same aforementioned entities, and the respective conclusions and recommendations should be published on the website of the regulatory authority.

Where recommendations have been formulated by the regulatory authority in the final report, these should establish a fixed period for the relevant utilities to provide information on their respective level of implementation.

When instances of non-compliance are detected which constitute an offence, the regulatory authority should start any respective proceedings, if they have the competency to do so, or send this information to the utility.

Regulation powers

The regulatory authority should be able to issue regulations with external effectiveness within the framework of their respective duties, for example with regard to the following matters:

- Tariffs, specifically defining the principles and rules to be adopted when establishing them, complementary to existing legislation.
- Quality of service, specifically defining the indicators, the minimum levels of quality, and any compensation payable in the case of non-compliance, supplementing existing legislation.
- Quality of drinking water, specifically defining the indicators and the minimum levels of quality, supplementing existing legislation.
- Regulatory procedures inherent to their relationship with the utilities, within the scope of their respective duties, thus specifically establishing the form and the period for carrying out their competences with regard to regulation.

Regulatory authorities should insure the participation of interested parties in the process of drawing up regulations with external effectiveness, under the terms laid down in the regulation for regulatory procedures, for example obligatory public consultation during the preparation stage.

Powers for the resolution of conflicts

The regulatory authority, within the scope of its respective duties, should be able to intervene in the resolution of disputes, specifically through conciliation, mediation and arbitration, between any utilities subject to its intervention or between these utilities and the uses of the services provided by them.

Conciliation is a process involving consensual, voluntary and private resolution of litigation, and involves a conciliator and the litigants. It involves face-to-face confidential sessions, in which only the final agreement may be known.

The conciliator, not party to the conflict, puts the litigants in contact and facilitates dialogue between them, without ever suggesting any proposals for consensus or even less issuing any binding decision. In this way, it only helps litigants to find their own path to building a solution, through the analysis of the weak and strong points of opposing positions, the advantages of reaching an agreement in which there is a concession from one party to the other, with such an agreement not being able to respect or meet all expectations.

As it is a consensual process, the parties therefore have control of the result and the terms of the process, to the extent that the resolution of the dispute in its final and definitive form always depends on their agreement.

Mediation is also is a process involving consensual, voluntary and private resolution of litigation, and which involves a mediator and litigants. It involves face-to-face confidential sessions, in which only the final agreement may be known.

However, the mediator, not party to the conflict, proactively presents a recommendation or proposal of agreement to the litigants produced by itself. The mediator does not have the power to issue a decision which is binding on the litigants, but it is only able to help them in their own path towards constructing a solution to the dispute, intervening so as to suggest the steps towards an agreement.

As this is a consensual process, the parties have relative control over the result of the terms of the process, although to a lesser extent than in conciliation, to the extent that despite the proposed solution coming from the mediator, the resolution of the conflict in its final and definitive form always depends on agreement between the parties.

Arbitration is a process of resolving conflicts through an adjudication process, which involves an arbitrator and the litigants.

The arbitrator, not part of the conflict, and based on law or according to judgements of equity, has the power to impose a decision made by it on the litigants which is binding on them.

As this is not consensual process, the parties do not have any control over the result of the terms of the process, to the extent that the resolution of the conflict in its final and definitive form always depends on the arbitrator. Agreement by the parties merely consists of submitting a conflict to arbitration, except when the law dispenses the utility with agreement to enforcing this service, being sufficient for the user to request arbitration.

Sanctionary powers

The regulatory authority should be given the competence to consider offences and levy the corresponding penalties and also any other applicable sanctions to offences arising from legislation or regulations the implementation or supervision of which falls within their remit. It can also levy penalties as a result of non-compliance with its own resolutions, under the terms envisaged in law.

Disrespect for binding instructions determined by the regulatory authority constitutes an offence, especially those of a preventative and cautionary nature, as is stopping or making access for the regulatory authority difficult with regard to facilities or documents relating to the carrying out of regulated activity. It is also an offence to not submit or submit false, inaccurate and incomplete information, requested by the regulatory authority or when the presentation of which is legally required.

The result of the levying of penalties and other compulsory financial sanctions should ideally not revert in favour of the regulatory authority, so as to ensure its independence of decision making regarding sanctionary aspects, or, if it does revert to them, the amount should clearly be earmarked for the sectors, for example expenses inherent to capacity building activities for the regulatory authorities.

The regulatory authority should publicly disseminate and identify offenders, describe the offences, the standards violated and the sanctions applied on their website for all the offences which fall under its administrative decision-making competence. In addition, it should be published in a daily newspaper of national scope and a local or regional periodical of the area of the head office of the offender.

4.5.6 Regulatory scope

In terms of the scope of the regulatory authority, it is necessary that the regulated activities be comprehensively defined, as well as the universe of regulated entities and the geographical area of the regulatory intervention, aspects which will be analysed below.

Regulated activities

The regulatory authority must have the regulated activities clearly defined, and there are of course various possible options.

It is for example possible to work in a way which is limited in terms of sectors, covering public drinking water supply, waste water management and solid waste management, that is, the three structural and irreplaceable services of modern societies. Their proximity and interdependence are grounds for joint regulation, without losing any necessary specialisation. Regulation may thus benefit from the potential to take advantage of the efficiencies of the closed urban water cycle, jointly regulating the public drinking water supply and waste water management, and the joint management of sub-products, such as sludge, jointly regulating waste water management and solid waste management. In this case it is necessary to make it clear if urban storm water management services and industrial waste services are included or not.

The so-called network activities such as energy services, telecommunications, transportation, water and waste, may be regulated separately or jointly, as a whole or in part. This option depends on various factors and the existing context, with there being no perfect model.

Universe of regulated entities

The regulatory authority should also have clearly defined the universe of regulated entities.

All utilities which provide water and waste services are considered, as mentioned previously, to fall under the scope of the regulator, regardless of whether the State or municipality owns the respective systems and the model of management adopted, namely whether this is a service directly provided, delegation of service or a service concession. Whenever the rights and obligations of the utility or the users are in question, then the entities that are the holders of the water and waste services should also be subject to the regulatory authority. This ensures the same level of protection to the user, regardless of the type of utility providing the service, in terms of access, quality and price.

Geographical area for regulatory intervention

The regulatory authority should have clearly defined the geographical area of its operation.

Regulatory intervention should be geographically extended preferably to the national level, although this may be regional in large countries. This enables a more global view of the sectors, the harmonisation of rules, procedures and interpretations in an extended territory, with the possibility of benchmarking a more significant number of utilities, reducing the risk of regulatory bias through the multiplicity of relationships and, of course, the rationalisation of regulatory resources, providing lower unit costs for users.

4.5.7 Public disclosure of accounts

With the regulatory authority carrying out the State's duties involving organisational, functional and financial independence, but without its directorate being subject to an electoral procedure, its legitimacy must be based, in addition to the proceduralisation of decisions involving the participation of interested parties (sounding out advisory bodies, public consultations and contradictory periods to allow opposing ideas), in mechanisms involving transparency, public accountability and control.

It should, therefore, publicly render accounts to political institutions, reporting regularly on its activity, particularly through relevant performance indicators.

Disclosing accounts to the legal power and the possibility of interested parties appealing its decisions also constitute important control mechanisms.

There cannot be independence without accountability in carrying out public posts. The public accountability is a basic demand of all public powers. A lack of such public provision may raise questions about the legitimacy of the independent regulatory authority.

The regulatory authority should therefore be subject to mechanisms of transparency and the accountability, particularly:

Plans and activity reports

The regulatory authority should draw up an annual activities plan, thus simultaneously ensuring the principles of transparency and desirable predictability for regulatory intervention, as well as an annual report of activities carried out, thus ensuring the principles of transparency of regulatory intervention and the dissemination of its actual activity.

Advisory body hearings

The regulatory authority should send an annual activities plan and its annual activities report for an opinion by its advisory body, which consists of representatives of all the sectors' stakeholders.

The regulatory authority should also accept, whenever it is so required to do so, requests for hearings to be run by its advisory body for the disclosure of information or clarifications concerning the activities of the regulatory authority.

Parliamentary hearings

The independence of the regulatory authority does not make it immune – in fact, rather the contrary - to parliamentary monitoring, and its chairman may be requested to appear before the respective parliamentary committees, so as to clarify or explain any regulatory initiative or measure. Parliamentary scrutiny at the committee level, which is less politicised and more technical, may be an appropriate counterpart in the absence of governmental control.

The regulatory authority should draw and annually send to parliament a summarised report on its respective activity concerning regulation and supervision. Indeed, the law should oblige the regulatory authority to draw up an annual plan of its regulatory activity, which should not be confused with the normal management activity report and the accounts of the authority. It should specifically be sent to government and the competent parliamentary commission and should also be publicly disseminated. This is one of the most visible means of accountability.

The regulatory authority should also accept, whenever it is so required to do so, requests for hearings to be run by parliament for the accountability, or clarifications concerning its activities, in addition to the obligatory annual report and the statement concerning the nomination of members to its board of management.

In this way the necessary scrutiny is substantiated, through the provision of accountability to parliament and consequently to society.

Transversal administrative and financial control

The regulatory authority should find itself, within the scope of financial responsibility and without prejudice to the activity of its own supervisory body, subject to the jurisdiction of general transversal control bodies for public activity, albeit with the changes required for the specific nature of those authorities.

This specifically includes the courts, (contentious dispute of its decisions, in the case of illegality), the court of auditors, regarding financial management, including the carrying out of financial responsibility, and other independent bodies with specific competences involving control of public power, such as, should they exist, the Ombudsman and the commission for the protection of personal data.

In effect, independence is only justified in relation to governments, utilities and other sectors' stakeholders.

Review of regulatory decisions

Questions concerning appeal, revision and the carrying out of decisions, orders or other measures legally susceptible to dispute which are taken by the regulatory authority may require a court specialised in competition, regulation and supervision. All the other acts of an authority of an administrative nature carried out by the bodies of the regulatory authority may alternatively fall under administrative jurisdiction.

Transparency and publicity

The most important guarantee of public responsibility is the very public nature of the exercise of power. The fundamental requirements for transparency include the dissemination of all relevant data, the prior public announcement of draft regulation and the most important regulatory measures, and procedural democracy, namely the intervention of all interested parties in the decision-making process and the justification of decisions taken.

Transparency is very important, not only as a tool of combating different practices of corruption, but also to prevent the camouflage of politically unpalatable information, for example water revenue being used for other purposes, or water quality incompliances being hidden from the public.

Given this, the regulatory authority should make available on its website all the important data for the sectors and their activity, particularly the composition of its statutory bodies, including biographical and bio data of the respective members, its internal organisation, statue, laws and regulations that concern the regulated sectors, as well as the main regulatory instruments and the activity of regulation. It should also make available its annual water and waste services reports, the summarised reports concerning its activity, its management instruments, particularly its activity plans, activity reports and approved budgets, for at least the last three years, and its budgets and approved accounts, also referring to at least the last three years.

Public procedural participation

The best defence against regulatory bias is the transparency of proceedings, the participation of all interested parties in the regulatory process, starting with the regulatory options adopted, and the basis for decisions, above all when these have been contested during the procedural stage.

Transparency and public participation go hand in hand, with strong public participation strengthening the position of the regulation authorities, rather than diminishing their power. Public participation provides for instance the opportunity for public understanding and support for needed investment which has an associated impact on water and waste tariffs.

The risk of regulatory bias often comes from utilities subject to regulation. One of the best antidotes to this is the institutional participation of the other interested party in the activity of regulatory authority, the corresponding users, participating for example in the advisory boards within the regulatory authority. Their intervention is decisive in all regulatory procedures, particularly those to do with the establishing of tariffs, the approval of regulations and the processing of user complaints.

Public participation can be at different levels from the least involved level of information provided with water bills, to the most direct involved level of communities on their water and waste services.

For public meetings to be effective, they need to be held on a regular basis to achieve trust and transparency and with clear information available. Sporadic meetings held only when there is something very controversial to discuss are liable to become confrontational and not so useful. Water consumer councils, functioning either as part of a regulator, or preferably wholly independent, provide an effective voice for the public.

In developed countries, with good water services, the uses may be satisfied with the understanding that there is full and good information available. They are likely to be satisfied that there are independent regulators and consumer associations, who will look after their interests and report any problems on due time. The situation depends critically on whether the uses are confident that there is full transparency, as any suspicion that something is being hidden will undermine the credibility of the authorities. However, in the developing countries, public participation may be more necessary to provide the incentive for improving services, and the means by which improvements must be achieved.

Regulatory impact assessment

Regulatory impact assessment (RIA) is a systematic process to identify and quantify costs and benefits resulting from the adoption of any relevant proposal of a regulatory nature which is being considered.

In fact, in seeking to have an effective regulation that constantly attains the public policy objective and leads to the implementation of efficient regulation which reaches these objectives at the least cost for all members of society, the regulatory authority should assess the regulatory impact beforehand, critically quantifying the positive and negative effects of alternatives to regulation, publicly providing an account of its results, and thus promoting regulation which is as efficient and effective as possible.

4.6 ORGANISATIONAL STRUCTURE OF THE REGULATORY AUTHORITY

4.6.1 Overview

Within the framework of good practices to be adopted in setting up or reorganising a regulatory authority, the present section describes a possible organisational structure suitable for the regulatory authority, which in general should have a management body, an advisory body and a supervisory body.

4.6.2 Management body

The robustness of the regulatory decision process requires a suitable governance model for decision taking within the regulatory management body, clear definition of the relations and the level of interference between this body and the members of the governments responsible for the sectors and provisions concerning conflicts of interest.

The management body should be collegiate and responsible for the definition and implementation of the activity of the regulatory authority, as well as for the management of its respective services, in conformity with the law and its bylaws.

It should be made up of an odd number of members, for example a chairman and two or four members, one of which may be designated vice-chairman. The chairman and the members of the management body should be appointed from among professionals of recognised repute, independence and technical and professional competence, in a transparent and credible manner, with the involvement of governments and parliament. They should be nominated for a period which is sufficiently long to ensure regulatory stability but sufficiently short to ensure desirable rejuvenation, for example, with a non-renewable mandate of between five to ten years.

Mechanisms regarding ineligibility, incompatibility and impediments are varied in nature, and relate to three distinct stages: firstly, precluding the selection of those who have compromising links or interests in bodies subject to the regulatory jurisdiction of the authority in question; secondly ban any conflict of interests of regulators with regulated utilities, for example forbidding them from holding shares in the companies in question; thirdly, not allowing the regulators to be able to join regulated services immediately after ceasing their mandate. Party political independence, particularly not being involved in electoral campaigns or other public party political activities, should also form part of the ethical requirements of the independent regulator.

The members of the management body should carry out their functions exclusively, not having any interests of a financial nature or holdings in bodies subject to intervention by the regulatory authority. After the period of their mandate and for a certain specified period, for example two years, the members of the management board must be forbidden from carrying out any role in or providing any service to entities formally subject to their regulation. The members of the management body are also subject to the general incompatibilities and impediments scheme established for holders of senior public posts.

In order to ensure the impartiality of the members of the management body, a member should not be nominated who, at the moment of being nominated or in the two years prior to this, is or has been a member of the managing body of the regulated utilities or is or has been employed or an on-going collaborator for the same entities, in a management or supervisory capacity in the same period, or even is or has been a member of the managing body of sectoral business associations.

The duties of the chairman of the management body are, besides calling and chairing the meetings of the body, guiding its work and promoting compliance with its respective decisions, coordinating its activity and its relations with this and other bodies and services of the regulatory authority, coordinating relations with governments, with other public bodies and with other holding and management bodies, request the convening of the advisory body to consider matters he considers important, carry out the duties delegated by the management body and any other duties provided for in the bylaw or in legislation.

The management body may take on itself the functions of regulatory decision or take on only the duties of supervision and strategic guidelines and delegate the regulatory decision functions to a chief executive officer.

The competences of the management board regarding regulation and supervision include the practice of any acts necessary for carrying out the duties of the regulatory authority, making decisions, issuing opinions, issuing binding instructions, requesting any precautionary measures, specifying inspection actions, auditing and supervision, exercising its administrative power, proposing public policies recommendations, drawing up draft legislation, approving regulations with external effectiveness as provided for in the law, issuing recommendations and codes of best practice, entering into cooperation and collaboration protocols, co-ordinating and carrying out the collection and dissemination of information and promoting research, innovation and the carrying out of studies.

The competencies of the internal management of the management board include administering the activity of the regulatory authority and its services, drawing up the annual activities plan and ensuring its respective implementation, monitoring and assessing, drawing up the budget proposal, under the terms of the applicable legislation. Other duties include carrying out any necessary budgetary alterations, drawing up the annual activities report and accounts, drawing up the body's balance sheet, under the terms of the applicable legislation, exercising powers related to the administration, management and disciplining of staff, approving the internal rules necessary for carrying out its duties, practising any other management acts which are shown to be necessary for the good functioning of the services, monitoring and systematically assessing the activity carried out and nominating representatives from the regulatory authority for external bodies.

Although the immovability of the members of the management body of the regulatory authority is one of the guarantees of its independence, an exception naturally has to be made in the event of serious offences. In these cases, and only after suitable investigation process, governments may dissolve the board of regulatory body, in the case of collective responsibility, or individually remove the offending members. In the final instance, in the event of repeated bad conduct, there is also the last resort involving the termination of the regulatory authority, since its independence is never absolute.

4.6.3 Advisory body

The regulatory authority should arrange its own advisory body to provide support in specifying its general areas of operation, thus ensuring the participation of representatives of the main interests involved in the water and waste services activities.

The advisory body should issue an opinion regarding the annual activities plan and report and its accounts, regarding the regulatory model and regarding other matters, consideration of which will be submitted to its management body. It must also present, on its own initiative, suggestions and proposals to the management body aimed at promoting the improvement of the sectors and the activities of the regulatory authority within the framework of its respective duties.

The advisory body, which may operate with specialised sections, may be chaired by a person of recognised merit, appointed in a transparent manner, preferably with the involvement of the government and parliament.

The advisory board may include representatives from all the important agents of the sectors, particularly:

- Environmental, water resources, public health, consumer protection and competition authorities at the national level and, if required, the regional level.
- State, municipal and private utilities, covering the various existing management models, namely direct, delegated and concessionary management, and covering both water and waste services.
- User associations at a national level.
- Associations representing economic activities.
- Significant technical and professional associations in the sectors.
- Non-governmental environmental organisations.

The advisory body should be independent in carrying out its duties, particularly from the management body, and not be subject to specific instructions or guidelines.

It constitutes a form of representation of all society and is essential to test the validity of the methodological principles of the regulatory instruments.

4.6.4 Supervisory body

The regulatory authority should arrange for its own supervisory body responsible for verifying the legality and efficiency of its financial and asset management. The supervisory body should be a statutory auditor or company of statutory auditors, without any situation which may create incompatibility. It is appointed, for example, for a period of three years, which may be renewable once.

The supervisory body should monitor and control the financial and asset management of the regulatory authority, consider and issue an opinion on its budget, its activities report and the annual accounts of the regulatory authority, confirm that the accounting has been executed well and in compliance with applicable provisions in terms of its budget, accounting and cash management, informing the management body of any anomaly which may be discovered. It shall draw up quarterly reports regarding its supervisory activity and an annual general report, and rule on matters within its competence that are submitted to it by the management body.

The supervisory body should be independent in carrying out its functions, and not be subject to specific instructions or guidelines, and be governed by the legal provisions respecting the performance of this type of activity.

4.6.5 Organisational model

The organisational and functional model of the regulatory authority should be based on a suitable organising structure, involving for example the management body, the operative areas of economic and financial analysis, water engineering, waste engineering, legal analysis and water quality, and also the transversal support areas of studies and projects, information technologies and administrative and financial areas, including secretarial services. This structure should be supplemented by the other advisory and supervisory bodies already mentioned.

The organisational model may for example be based on a functioning matrix, in which the operative areas and the support areas of a vertical nature, besides the specific areas, may allow for the constitution of horizontal work teams, bringing together the necessary valences of the various sectors for the resolution of problems which in general are interdisciplinary. This is particularly the case in many matters involving regulation, with joint intervention involving economic and financial analysis, engineering, legal analysis, quality of drinking water and study and projects, and the development of the information system.

Alternatively there can be an organisational model centred on knowledge, and a functional model may be used, for example through structural regulation, legal and contractual regulation, economic regulation, quality of service regulation, drinking water quality regulation and regulation of user interaction.

A recipient based model may also be utilised, for example with departments focused on direct management utilities, delegated management utilities and concession management utilities.

It is up to the management body to opt for the model that it considers most suitable in each context.

4.7 RESOURCES OF THE REGULATORY AUTHORITY

4.7.1 Overview

Effective regulation, in addition to independence, requires resources, with it being essential to ensure suitable human, budgetary, financial, technological and physical resources.

This section describes the main required resources within a framework of good practices to be adopted when setting up or reorganising a regulatory authority.

4.7.2 Human resources

The regulatory authority should have available operational services providing technical and administrative support which are essential for the effective operational performance of its duties, without prejudice to the possibility of resorting to the outsourcing of services to external entities.

As such, it should make available the capacity for self-organisation as regards staff, careers, administrative posts, salaries and recruitment, naturally falling within general rules. This ensures greater freedom in terms of modifying its structure and for the development of the sectors and the needs of regulatory intervention. This autonomy of human resource management constitutes a further guarantee of its independence, and should not for example be limited by the national budget law, but rather constituting the rule within the scope of the legal system of independent administrative bodies.

Human resources should have suitable professional training and experience and in-depth knowledge of regulation. There should be as small a staff as possible, but with specific training in the working areas in which they are involved. They must have experience and be knowledgeable of the sectors, with the university degree level but preferably with additional training, with the ability to co-ordinate and link work within an outsourcing regime. The regulatory authority should invest in ongoing and specialised human resources training, so that it is kept continually updated.

It should be possible to contract external qualified consultants and independent auditors wherever necessary, particularly to respond to peak periods in the regulatory cycles, which do not always justify full-time staff. An example is the period of the annual quality of service audits, which in general require a large additional team effort.

The conditions for the recruitment and selection of staff should observe the general principles of publishing employment vacancies, equality of conditions and opportunities for candidates, application of methods and rigorous objective criteria for evaluation and selection, ensure there are no incompatibilities and there are sound justifications for all decisions taken. In fact, a clear system of incompatibilities should be specified so as to ensure administrators and employees perform their roles impartially.

The members of corporate bodies of the regulatory authority, as well as staff and, with the necessary applications, interns, external consultants and other service providers should remain subject to the duty of professional confidentiality regarding facts and documents, knowledge of which stems exclusively from the functions they carry out. They may not disseminate or use the information obtained except in the strict exercise of their roles.

It should be noted that the technical staff of the regulators are very much sought after by the regulated sectors. Good incentives are needed both to contract them, as well as to keep them. It is an error to compare the staff of the regulators to public entities as a whole, in so far as they have to be extremely specialised and adopt a position of major impartiality. Since the utilities often have a high technical and legal skills level to oppose the regulatory authority, effective regulation is not possible without highly competent human resources.

4.7.3 Financial resources

The regulatory authority should have the necessary financial resources available to effectively carry out its activities and be able to be financially autonomous.

Its financial resources should come exclusively from the collection of levies for the regulation services provided to the utilities, supported by the users and not by the taxpayers, thus ensuring its financial independence and objectivity regarding political power and reinforcing confidence in the regulator. These charges should constitute one of the criteria for the establishing of the tariffs to be practised by the utilities, that is in the final instance regulation must be paid for by the users.

The regulatory authority should have budgetary autonomy within the rules of auditable economy, efficiency and effectiveness by the competent bodies, which ensures suitable conditions of independence in carrying out its functions, plus avoiding or reducing the blocking of the purchases of services and goods necessary for its activities. This constitutes a guarantee of its independence in the use of its revenues, without interference or obstruction, subject however to existing general controls.

The impact of the regulatory cost for the user should be sufficiently low, in terms of its contributory effects to the tariff load. It should in general be less than one percent of the average tariff. The values for regulation charges should just be based on specific benefits to society, considering the effects in economic, environmental and social terms. It is important that the users have the perception that the regulation costs which they bear through the tariffs are low and clearly compensated for by the benefits they obtain from that same regulation.

4.7.4 Physical and technological resources

The infrastructure resources of the regulatory authority should include suitable facilities in terms of the number of staff, and should be functional and robust, but never ostentatious. Facilities granted by one of the sectors' stakeholders, for example the government, should never be utilised, since this would indicate a form of bias.

Given the large number of documents and information which results from the extremely assiduous relationship with the utilities, which may have a high number, the regulatory authority should be endowed with high-capacity information technology infrastructure with a high level of reliability and sufficient redundancy.

The regulatory authority should implement an information system capable of managing a large quantity of information, which enables the exchange of information with all the utilities and other stakeholders more quickly, simply and effectively, enabling information reporting by all the utilities of regulated services, as well as access to a set of important information. As will be detailed below, the information system modules for the regulatory authority should provide support for the various regulation components, particularly legal and contractual regulation, economic regulation, regulation of the quality of service, regulation of the quality of drinking water, and the regulation of the user interface, as well as the website accessible to the general public where the most important information on the water and waste sectors should be disseminated. There should also be internal models of the information system covering for example document management, process management, activity control, administrative management, financial management and archive management.

4.8 SUMMARY

This chapter covered aspects such as the need for regulation of the services, the objectives of regulation, the regulatory mission and mandate, characteristics of the regulatory authority, the organisational structure of the regulatory authority and its human, financial, physical and technical resources.

In the next chapter the integrated regulatory approach that is proposed will be described, as a tool for the regulation of the water and waste services.

Chapter 5

Integrated regulatory approach

5.1 INTRODUCTORY NOTE

Implementing a regulation model for the public drinking water supply, waste water management and solid waste management services should be carried out with much reflection, thus contributing to the improvement of all aspects regarding the services and not just in a partial manner, seeking in this way to find the ideal global solution.

This chapter presents a model of regulation based on an integrated approach for water and waste services (model RITA-ERSAR), seeking to find a suitable balance regarding the various aspects involved.

5.2 INTEGRATED APPROACH TO REGULATION

When setting up a regulator for water and waste services, it is essential to define a clear and effective regulation model, which is rational with regard to the local context of these services, while also benefiting from the experience of regulation in other activity sectors and, of course, from international experience. This clarity is indispensable so that all the stakeholders involved in the sectors, especially the utilities, know the rules of the regulation model in advance and can decide their positioning with greater security.

The regulation model to be adopted may of course vary according to the services in question (water services, waste services or others) and with regard to the actual context in which the services will be developed, taking into account, in an integrated manner, technical, economic, legal, environmental, social and ethical aspects, and whether this is to be implemented within a short, medium, or long-term perspective, with stable rules independence, capacity, impartiality and transparency. It should be clear, simple and practical for the users.

Regulation should be applied with appropriate force, variable in terms of the context and the powers which are attributed to the regulator. Regulation can be, for example, more impositional, such as is the case with the United Kingdom, more cooperative, as is the case with Portugal, or even self-regulating, as in Germany and the Netherlands.

The conceptualisation of the regulation model therefore depends on the existing context, that is, the real situation for the water and waste services and the surrounding political, economic, social and environmental context.

It is essential to understand that there are no universal solutions and, for each reality, whether this be a country or region, the most suitable regulatory model should be designed and then the details added. It is clearly a risk to adopt a regulatory model that has been imported and which has not been adapted to the reality of the country or region.

It is however very important that that conceptualisation of the regulation model be carried out in an integrated and holistic manner, to take into account problems which are both specific and global in nature and seeking an integrated regulatory approach for the water and waste services, which can resolve the various problems separately but which can also find the optimal global solution, through a suitable balancing of the various aspects involved. The model should thus be applied through distinct components but also ensure there is close linking and in this way enhance the synergies between these same components. For example, economic regulation should be an enhancing factor but also benefit from quality of service regulation, and drinking water quality regulation should influence economic regulation and also benefit from it. This connection should be seen between practically all the components of the regulatory model.

Based on the practical experience of around 12 years of regulation in Portugal[1], it is described in this book a model of regulation with an integrated approach (designated as RITA-ERSAR) for the water and waste services. This model aggregates the various strategic, technical, economic, environmental and social components, seeking an appropriate balance of the various perspectives at stake. It regulates both the sectors as a whole and also the utilities individually, and is simultaneously applied in a relatively similar manner to the different public services, specifically drinking water supply, waste water management and solid waste management, with suitable adaptations, seeking to find the optimal global solution.

This integrated regulatory approach for the water and waste services is implemented through two main levels of intervention, as described below:

- A first level, aimed generically at the sectors as a whole, designated as structural regulation of the sectors, which involves a contribution towards the better organisation of the sectors, through the clarification of its operational rules, the drawing up and regular dissemination of information regarding the sectors, and capacity building and innovation for the sectors. The regulator is not focused on any utility in particular, but on the sectors as a whole, helping to create organisation, rules and tools for its good functioning. It therefore corresponds to macro regulatory intervention.
- The second level, designated as behavioural regulation of the utilities, consists of its legal and contractual monitoring throughout the life-cycle, economic regulation, quality of service regulation, drinking water quality regulation and user interface regulation. In contrast to structural regulation, the regulator is focused here on each of the utilities operating in these sectors. It therefore, in a supplementary manner to the previous level, corresponds to micro regulatory intervention, multiplied by the number of regulated utilities.

The need for effectiveness in regulating natural and legal monopolies leads to the supplementary utilisation of the structural regulation of the sectors and the behavioural regulation of the utilities. Disconnected utilisation is necessarily less effective than this supplementary utilisation.

In the case of the public drinking supply of water, waste water management and solid waste management services, as they remain relatively static through time, with gradual alteration of market and technological conditions, there tends to be a prevalence of behavioural regulation of the utilities over the structural regulation of the sectors. However, in periods with more pronounced revisions of public policies, there

---

[1]In Portugal, over the last decade, ERSAR has designed and implemented a regulation model adapted to its national reality, specifying the corresponding procedures and creating the technological tools needed, which it has applied to around five hundred utilities.

then tends to be a reverse prevalence of structural regulation of the sectors over behavioural regulation of the utilities. The intervention of the regulator should of course be adapted to accompany such trends.

An analysis of the structural regulation of the sectors and the behavioural regulation of the utilities will now be carried out in greater detail.

5.3 STRUCTURAL REGULATION OF THE SECTORS

5.3.1 Overview

Structural regulation of the sectors is focused on the sectors as a whole and should contribute to its better organisation and to help to clarify its rules, such as restrictions on the entry of utilities into the market and functional separation measures, which define which entities or types of entity can provide these services. Structural regulation also includes a set of measures to consolidate and modernise the sectors, both by making information available and enabling stakeholder capacity building. This regulation is a form of direct control over the surrounding and indirect context concerning the utilities, reducing or eliminating the possibility of undesirable behaviour. Within a preventative logic, it strongly conditions the form, content and nature of behavioural regulation, so much so that it needs to be supplemented.

Any definitions or alterations of public policies for the sectors should necessarily be accompanied by the regulator which, while of course not possessing the competency for any such definition, as these are decisions of a political nature, should, however, help to substantiate such choices, particularly in order to guarantee the protection of users' interests and safeguard the utilities' legitimate interests, as well as assessing the level of acceptable risk for society.

As regards the structural regulation of the sectors, the regulator should contribute towards:

- organisation of the sectors;
- legislation of the sectors;
- information of the sectors;
- capacity building of the sectors.

Each one of these structural regulation components will now be briefly described.

5.3.2 Regulatory contribution to the organisation of the sectors

In this structural regulation component, the regulator should contribute to the formulation of better public policies, towards their rationalisation and the resolution of any malfunctions regarding the regulated services and towards the organisation of the sectors, promoting for example an increase in the efficiency and effectiveness of the water and waste services and the search for economies of scale, scope and process.

It should afterwards monitor the national strategies adopted for the sectors, by accompanying their implementation and regularly reporting on any evolution or constraints.

5.3.3 Regulatory contribution to the legislation of the sectors

In this structural regulation component, the regulators should draw up proposals for new legislation or the alteration of existing legislation, for example at the level of the legal framework governing the systems, technical legislation regarding the water and waste services and the legislation governing regulation.

In this way it should contribute towards a clarification of the rules for the provision of these services through proposed legislation and the issuing of regulations and recommendations.

It should afterwards monitor the application of legislation in force and those regulations and recommendations, assessing their effectiveness and the need for any improvements or replacements.

5.3.4 Regulatory contribution to the information of the sectors

In this component of structural regulation, the regulator should make available and regularly disclose thorough and accessible information to all sectors' stakeholders, through the coordination and carrying out of the collection, validation, processing and disclosure of information regarding the sectors and the respective utilities, as well as making that information available and the subsequent increased public interest in consulting it.

It should thus contribute to consolidating an actual culture of concise and credible information which can be easily interpreted by all, extendable to all utilities, regardless of the forms of management adopted for the provision of such services. It should enable suitable knowledge to be made available, based on the information obtained from the numerous data created within the sectors, thus ensuring the fundamental right of access to information of all users and society in general.

5.3.5 Regulatory contribution to the capacity building of the sectors

In this component of structural regulation, the regulator should provide technical support to the utilities through the production of technical publications in partnership with knowledge centres, the direct and indirect promotion of seminars and conferences, support for events by third parties, conducting opinion studies and promoting research and development in the sectors, thus motivating the academic sector in this field. It should also make itself available to answer the various questions put to it by the different sectors' stakeholders.

In this way it can contribute to an improvement in the technical capacity building of the utilities and to encouraging the consolidation of the national business sectors.

5.4 BEHAVIOURAL REGULATION OF THE UTILITIES

5.4.1 Overview

At the same time, the strategy of the regulator should also include the behavioural regulation of the utilities in their actions in markets subject to regulation, with regard to the legal and contractual, economic, quality of service, drinking water quality and user interface aspects, which will be described below.

These regulation components are strengthened through performance assessment and benchmarking with the results of other similar utilities operating in different geographical areas. These mechanisms should adopt a logic which is pedagogic and appreciative of worth, benefiting the utility in some manner in terms of its performance regarding the average performance of all utilities. To achieve this it is necessary that the regulator receives information from the utilities in the form of data so that previously defined performance indicators can be calculated and, after their validation, a comparative analysis is carried out with the historical records of the actual utility, so as to see the evolution through time of different management aspects, and compare with other similar utilities. This allows, in particular, performance levels to be defined and references to be established to enable the realistic setting of new efficiency targets. The results of this comparison should be publicly made available, in so far as this pressures the utilities in terms of their efficiency, as they naturally do not want to be seen in an unfavourable light, and so it implements a fundamental right available to all users.

As regards the behavioural regulation of the utilities providing water and waste services, the regulator should carry out:

- legal and contractual regulation;
- economic regulation;
- quality of service regulation;

- drinking water quality regulation;
- user interface regulation.

Each one of these behavioural regulation components will now be briefly described.

5.4.2 Legal and contractual regulation

In this component of behavioural regulation the regulator should ensure the legal and contractual monitoring of the utilities throughout their life-cycle, specifically through analysing the tendering and contracting processes, contract modifications, contract terminations, and reconfigurations and mergers of systems, accompanying the carrying out of contracts and intervening where necessary in reconciliation activities between parties.

Legal and behavioural regulation should thus contribute to ensuring public interest and legality.

5.4.3 Economic regulation

In this component of behavioural regulation, the regulator should ensure the economic regulation of the utilities, thus promoting the regulation of prices to ensure efficient tariffs which are socially acceptable to users without prejudice to the necessary economic and financial sustainability of the utilities, within an environment of efficiency and effectiveness in the provision of their service. Economic regulation also includes the assessment of investments to be undertaken by the utilities.

As monopoly prices tend to be higher than those resulting from competing markets, obtaining lower prices that will allow the economic and financial viability of the utilities and will correspond to a fairer system for users requires major intervention from the regulator.

Economic regulation should thus contribute to promoting the economic and financial sustainability of those utilities, without prejudice to the economic accessibility to the services for the users.

5.4.4 Quality of service regulation

In this component of behavioural regulation the regulator should ensure the regulation of the quality of service provided to users by the utilities, assessing their performance and comparing the utilities among themselves, through the application of a suitable selection of performance indicators, so as to promote effectiveness and efficiency, which represents an improvement in their levels of service.

Quality of service regulation is a way of regulating performances, which is inseparable from economic regulation. It constrains the permitted performances of the utilities regarding the quality of service they provide to users, steering the utilities in terms of effectiveness and efficiency, and embodying in this way a basic right of users.

This should therefore contribute to promoting an improvement in the levels of service provided to users.

5.4.5 Drinking water quality regulation

In this component of behavioural regulation the regulator should ensure drinking water quality regulation, assessing the quality of water supplied to the users, comparing the utilities among themselves and monitoring any non-compliances in real-time.

As the quality of drinking water is an aspect of quality of service, and as it has a major interaction with economic regulation, there is a clear rationale for it to be regulated by the same body, although it is not absolutely necessary to follow this model and this is not the case in various countries.

It should thus contribute to promoting an improvement in water quality and public health.

5.4.6 User interface regulation

In this component of behavioural regulation the regulator should ensure compliance by the utilities with consumer protection legislation and, in particular, undertake an analysis of any complaints and promote their resolution between users and the service provision utilities.

It should also foster the participation of service users, creating advisory mechanisms and disseminating information.

5.5 RITA-ERSAR REGULATION MODEL

The following is a graphic representation of the proposed RITA-ERSAR regulation model (Figure 5.1), which has been adopted by ERSAR, the Portuguese regulator, for more than a decade, corresponding to an integrated regulatory approach for the water and waste services.

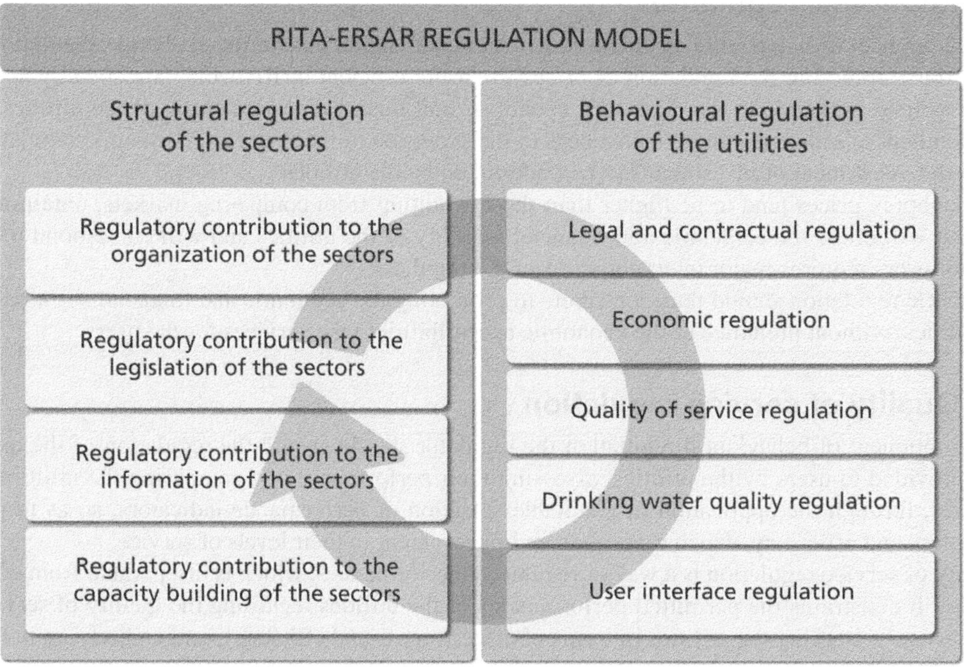

Figure 5.1 RITA-ERSAR regulation model.

As mentioned above, this model is based on two major levels of intervention: structural regulation of the sectors and behavioural regulation of the utilities. The components of the structural regulation of the sectors consist of contributions towards the organisation, legislation, information and capacity building of the sectors. The components of the behavioural regulation of the utilities consist of legal and contractual regulation, economic regulation, quality of service regulation, drinking water quality regulation and user interface regulation.

All these components must be perfectly linked to each other, so as to form a coherent and integrated model, the effectiveness of which depend from the synergies obtained between these components.

5.6 SUMMARY

The present chapter described the integrated regulatory approach being proposed as an instrument for the regulation of the water and waste services, describing the need for regulation, the goals and principles of regulation and the approach adopted.

The following chapters will describe both separately and in detail the different components of this model which is based on an integrated regulatory approach (model RITA-ERSAR), both in terms of its structural regulation component for the sectors, and the behavioural regulation of the utilities.

Chapter 6

Regulatory contribution to the organisation of the sectors

6.1 INTRODUCTORY NOTE

As part of this integrated approach (model RITA-ERSAR), this chapter describes in more detail the contribution made by regulations to the organisation of the sectors, one of the components of the regulation of public drinking water supply, waste water management and solid waste management services.

6.2 REGULATORY GOALS

It is naturally up to governments and not regulators to define public policies on water and waste services and establish the organisation of the sectors.

Although it is not the regulator's direct responsibility, the definition of good policies may benefit significantly from information, analyses and studies provided by the regulator, which thereby plays an important role in assisting the political decision-makers in supporting good decisions on public policies.

Furthermore, the successful implementation of these public policies for water and waste services depends on the ability to manage implementation of all the components at more or less the same time in an effective, overall, integrated approach. Some of these components depend on action by the regulator, with the structural regulation of the sectors contributing to the organisation, regulation, information and capacity building of the sectors, and behavioural regulation of the utilities, conducting legal and contractual regulation, economic regulation, quality of service regulation, regulation of quality of water for human consumption and regulation of the user interface.

Finally, the regulator should be responsible for monitoring these policies. It should be done regularly, independently and factually and identify their successes and failures, their causes and any measures that the government can take to better safeguard the national interest, comprising users' interest and the economic viability and legitimate interests of the utilities and other agents of the sectors.

The regulatory contribution to the organisation of the sectors therefore helps to ensure fulfilment of public service obligations, defined in 2.2, in terms of universal access to services, suitability of the quantity and quality of services, continuity of services and structural and operating efficiency of services.

It also helps achieve the goals of public policy, defined in 3.3, including appropriate strategic plans for the sectors, definition of the legislative framework, definition of the institutional framework, definition of the governance of the services, definition of access targets and quality of service goals, definition of tariff policy, provision and management of financial resources, construction of infrastructure and better structural and operating efficiency. It also helps with capacity building of human resources, promotion of research and development, development of the business world, introduction of competition, protection, awareness and participation of users and provision of information.

6.3 REGULATORY ACTIVITIES AND PROCEDURES

The main activities in this regulatory component are collaboration in formulating public policies, monitoring implementation of these public policies, namely the development of the sectors, periodically reporting the results of this monitoring, making improvements in the sectors and proposing measures for the rationalisation and resolution of malfunctions.

The regulator must collaborate indirectly in the formulation of national strategies by conducting permanent analyses and studies that support good decisions on public policies by the government. These analyses and studies should be based on information gathered in these components: legal and contractual, economic regulation, quality of service regulation, quality of water for human consumption regulation and interaction with users regulation.

It can thereby help to clarify aspects such as restrictions on the entry of utilities into the market and functional separation measures, which define which entities or types of entity that can provide these services.

It should also contribute to an appropriate level of aggregation of utilities by geographic unit, seeking economies of scale, an appropriate level of aggregation by type of service or market, seeking economies of scope, and an appropriate level of aggregation of stages in the production system, seeking economies of process.

The regulator must then regularly monitor implementation of these public policies and the development of the sectors, independently and factually, identify their successes and failures, their causes and any measures that the government can take.

This monitoring is essential for the timely identification and correction of deviations from the public policy goals resulting from structural aspects of the sectors. It also combines the results of performance in each of the regulated entities to provide a clearer perception of aspects that influence the results of policies in more detail.

The regulator must conduct periodic assessments and identify needs based on the experience acquired in the contribution to better organisation of the sectors. It must propose or promote when necessary implementation of rationalisation measures and resolution of dysfunctions to support the reassessment of public policies and, consequently, the improvement of the performance of the sectors and the utilities' activity.

These are generally unscheduled regulatory activities that the regulator performs on the basis of strategic plans for public policies and any casuistic needs that arise. Regular monitoring of the implementation of public policies must, however, be annual in order to regularly inform the government and the different agents on the development of the sectors, from the regulator point of view.

The regulator must follow a specific rational, clear procedure in its contribution to the organisation of sectors, which must be set out in a set of rules for regulatory procedures, ensuring respect for the principles of legality, need, clarity, participation and transparency.

There follows an example of a regulatory procedure that the regulator can follow. It is divided into stages:

- The regulator must constantly perform analyses and studies of the sectors and disclose their results, especially to the government, to support decisions on public policies for which the political powers are responsible.

- The government should consult the regulator prior to any strategic definitions or alterations for the sectors, in order to guarantee protection of users' interests and safeguard the utilities' legitimate interests.
- The regulator must regularly monitor implementation of these public policies and the development of the sectors, independently and factually, identify their successes and failures, their causes and any measures that the government can take.
- The regulator must publish the results of this regulatory activity, for example in the sectors annual report, in terms of regular monitoring of implementation of these public policies and the development of the sectors.
- Whenever opportune, the regulator must propose measures for the rationalisation and resolution of dysfunctions in the sectors, to assist in the reassessment of public policies by proposing improvements as a result of this ongoing monitoring or suggesting alterations due to changes of the context, requiring urgent measures by the government.

Figure 6.1 shows the different stages of the procedure of contribution to better organisation of the sectors.

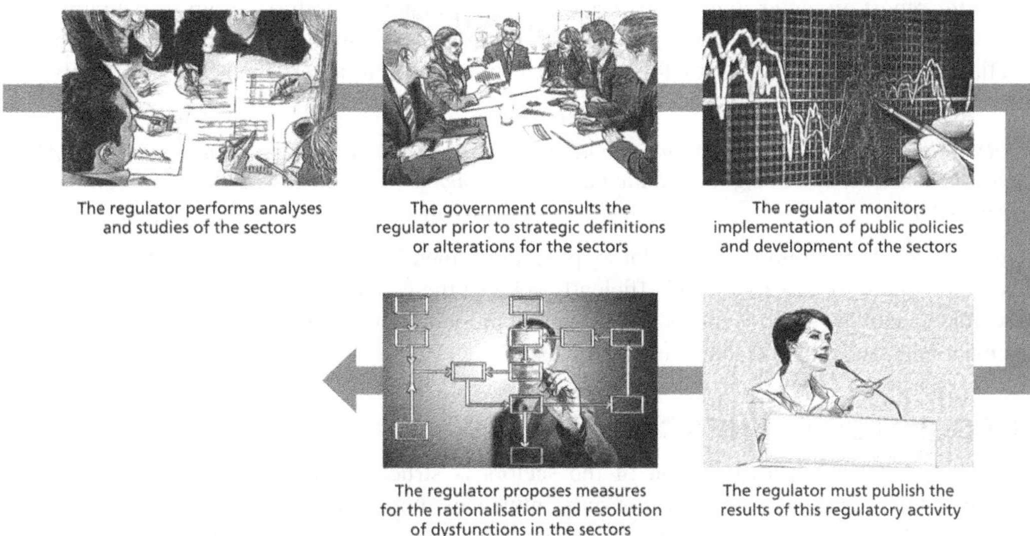

Figure 6.1 Regulatory contribution to organisation of the sectors.

6.4 REGULATORY INSTRUMENTS

The regulator must promote, develop or use the most appropriate instruments for its contribution to the organisation of the sectors. There follow some examples of regulatory instruments belonging to the regulator and external to the sectors:

- *Strategic sectoral plans and national programmes*: For this regulatory component, the regulator must naturally take into account medium-term strategic plans for water and waste services, even though these documents are not its responsibility and are usually distinct because the two subsectors have different characteristics. These plans embody the respective public policies. Furthermore, government programmes, political documents that are approved by the government and approved by parliament and apply to all sectors of activity and represent the country's public policies, must also be taken into account in matters with a direct and indirect impact on the water and waste services sectors. The regulator must operate within the framework of public policies defined for the country by legally elected political bodies, without prejudice to its independence. These plans and programmes must naturally be an essential reference for regulatory intervention in the contribution to the organisation of the sectors.

Strategic plans for the water and waste service sectors in Portugal have existed for the last 20 years. The last one, PEAASAR II, was approved by Ordinance 2339/2007 of 28 December and set the goals for 2007–2013. It proposed measures for optimising bulk and retail services and optimising the sectors environmental performance. It also clarified the role of private enterprise and created spaces for the affirmation and consolidation of a sustainable, competitive economic activity suited to the reality in Portugal. A new strategic plan for 2014–2020 is currently at the approval stage.

The Strategic Plan for Solid Waste 2007–2016 (PERSU II) was approved by Ministerial Order 187/2007 of 12 February. A new strategic plan for 2014–2020 is currently at the approval stage.

- *Annual report on water and waste services*: The regulator must draft an up-to-date annual report on water and waste services with the situation and development in the sectors for everyone needing reliable information in order to define policies or strategies to assess the services actually provided to users.

  ERSAR has published an annual report on water and waste services in Portugal since 2004 (RASARP). It includes a section describing the situation and development in the sectors and is available on its website (www.ersar.pt).

Thus, regulatory activity follows the regulatory procedures mentioned and these regulatory instruments and enables the regulator to effectively and efficiently achieve the goals of its contribution to the organisation of the sectors, one of the components of the regulation model for public drinking water supply, waste water management and solid waste management services.

6.5 REGULATORY SYNERGIES

The contribution to better organisation of the sectors is structural and articulates closely with other components of the regulation model. It not only influences those components but also benefits from the information gathered in economic regulation, regulation of legal and contractual compliance, quality of service, quality of water for human consumption and interaction with users, which may lead to the reorganisation of the sectors.

6.6 SUMMARY

This chapter gave a detailed description of one of the components of the proposed regulatory approaches within the framework of structural regulation, the contribution to the organisation of the sectors, including goals, activities, procedures, instruments and synergies.

The next chapter describes another of the components of the proposed regulatory approach, also in the framework of structural regulation, the contribution to the legislation of the sectors.

Chapter 7

Regulatory contribution to the legislation of the sectors

7.1 INTRODUCTORY NOTE

As part of the integrated (model RITA-ERSAR) regulatory approach this chapter gives a more detailed description of the regulatory contribution to legislation of the sectors, one of the components of the regulation model for the public services of drinking water supply, waste water management and solid waste management.

7.2 REGULATORY GOALS

The goal of this component of the regulatory model must be to help clarify the sectors rules of operation, which is naturally an essential aspect for the proper provision of these services, via instruments with varying degrees of obligation, specifically legislation, regulations and simple recommendations.

A sector that does not have clear operating rules that define the entities involved and their rights and obligations prevents these essential public services from being provided efficiently and effectively.

This component of the regulatory model thereby helps to achieve the public policy goals, defined in 3.3, with an appropriate legislative framework for the public services of the drinking water supply, waste water management and solid waste management to the population.

7.3 REGULATORY ACTIVITIES AND PROCEDURES

The main activities in this regulatory component are drawing up preliminary drafts of legislation, rules and recommendations and making improvements in the sectors:

- *Preliminary drafts of legislation*: When appropriate, the regulator must promote the preparation of preliminary drafts of laws to be submitted to the government. This should involve prior consultation of agents of the sectors, giving them the opportunity to give an opinion, via their advisory boards and public consultations. This support role from the regulator to the government and in parliament is very important. It would not be acceptable this task to be done by one of the other agents of the sectors, necessarily an interested party.

- *Regulations*: When appropriate, the regulator must promote the preparation regulations with external application as a complement to legislation produced by the government or parliament, at least on matters of tariffs, quality of service, quality of water for human consumption, commercial relations and regulatory procedures, within the scope of their competences.
- *Recommendations*: When appropriate, the regulator must promote the preparation of recommendations on, for example, specific aspects in the design, execution, management, operation and monitoring of water and waste systems. These recommendations, which are guidelines and can be implemented voluntarily by the utilities and other agents of the sectors, must be aimed at completing legislation and regulations, clarifying pending issues, specifying procedures and assisting the utilities in their activity.
- *Making improvements in the sectors*: Based on experience acquired in continuous monitoring of compliance with legislation and regulations and acceptance of recommendations, the regulator must periodically reassess and identify the need and, if necessary, directly or indirectly promote implementation of amendments to legislation, regulations or recommendations in order to improve the operation of the sectors and the activity of the utilities.

These are casuistic regulatory activities that are necessarily scheduled and performed by the regulator as needed.

For the preparation of preliminary drafts of legislation, the regulator must follow a rational, clear, specific procedure set out in the rules on regulatory procedures, ensuring respect for the principles of legality, necessity, clarity, participation and publicity.

There follows an example of a regulatory procedure that the regulator can follow when drawing up preliminary drafts of legislation. It consists of a number of stages (Figure 7.1):

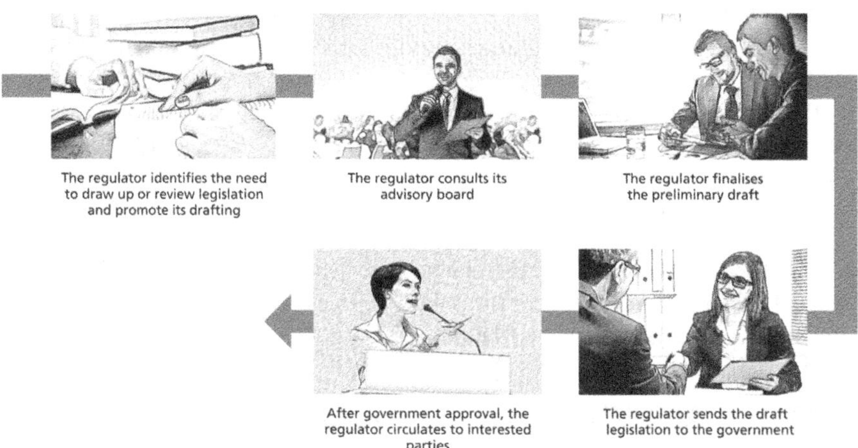

Figure 7.1 Regulatory contribution to the legislation of the sectors.

- The regulator must periodically identify the need to draw up or review legislation and, provided that it has been accepted by the government, promote its drafting or review, in the form of a preliminary draft.
- The regulator must consult its advisory board on the draft legislation, wait for its opinion and then review it on the basis of this opinion, thereby guaranteeing the participation of representatives of the sectors' main agents.

- When appropriate, the regulator must also submit the draft legislation to the interested entities in the form of a public consultation for them to make comments and suggestions by a preset minimum time, and then conduct a review on the basis of these comments.
- The regulator's management body must finalise the preliminary draft by attaching a report summarising the contributions made and justifying the decisions made.
- The regulator must send the draft legislation to the government so that it can begin the legislative process.
- After government approval and publication, the regulator must post the legislation on its website to ensure that it is circulated to all directly interested parties, without prejudice to other suitable forms of publication.

When drafting regulations with external efficacy and recommendations, the regulator must also follow a rational, clear, specific procedure set out in the rules of regulatory procedures, ensuring respect for the principles of legality, necessity, clarity, participation and publicity.

There follows an example of a regulatory procedure that could be followed by the regulator for drafting regulations with external efficacy and recommendations. It consists of a number of stages (Figure 7.2):

- The regulator must periodically identify the need to draw up or review regulations with external efficacy and recommendations and promote their preparation or review, in the form of a draft.
- The regulator must consult its advisory board on the draft regulations and recommendations, wait for its opinion and then review them on the basis of this opinion, thereby guaranteeing the participation of representatives of the sector's main agents.
- When appropriate, the regulator must also submit the draft regulations and recommendations to the interested entities in the form of a public consultation for them to make comments and suggestions by a preset minimum time and then conduct a review on the basis of these comments.
- The regulator's management body must finalise the draft regulations and recommendations by attaching a report summarising the contributions made and justifying the decisions made.
- The regulator must publish the regulation in the official journal and post the regulation or recommendation on its website to ensure that it is circulated to all directly interested parties, without prejudice to other suitable forms of publication.

Figure 7.2 Regulatory contribution to the regulations of the sectors.

7.4 REGULATORY INSTRUMENTS

The regulator must promote, develop or use the most appropriate instruments for its contribution to the clarification of the rules of the sectors. There follow some examples of regulatory instruments:

- *Legal framework of services*: For this regulatory component, the regulator must ensure that there is appropriate legislation on the services' legal framework. This legislation must define the overall legal framework of all the types of systems, harmonise the frameworks applicable to the different direct, delegated and concessioned management models to protect end users, clarify the hiring of services with users and ensure equality and transparency in access to and performance of the activity and in contractual relations. It must also ensure correct protection and information for the users of these services and prevent possible abuses of exclusive rights in terms of the quality guarantee and control of the public services provided and the supervision and control of prices, which is essential if there is a monopoly. It must also include a penalties regime.

  In Portugal, under the law on the powers and competences of the local authorities (Law 159/99 of 14 September), it is the municipalities' responsibility to ensure the supply of water, treatment of waste water and solid waste management services to the population via the urban systems.

  The Sector Delimitation Law (Law 88-A/97 of 25 July, which repeats the wording introduced in 1993 in this regard) introduced multi-municipal water and waste systems and defined them as systems that serve at least two municipalities and require a predominant investment from the State for reasons of national interest (considering all others to be municipal systems). This distinction is relevant in determining the form of access by private companies to the management of water and waste services. In multi-municipal systems it is only allowed via a minority holding in the concessionaires, while in municipal systems the concessionaires are allowed to own all the share capital.

  The fundamental rules on management of multi-municipal systems are set out in Decree-Law 379/93 of 5 November, which was later developed by Decree-Law 294/94 of 16 November, Decree-Law 319/94 of 24 December, and Decree-Law 162/96 of 4 September, all with the latest wording set out in Decree-Law 195/2009 of 20 August, which republished them. These laws approved the bases for the multi-municipal concessions for solid waste, public water supply and waste water treatment systems, respectively. The current thirty or so multi-municipal systems were created by decree-law under this framework followed by a concession contract between the State and concessionaires set up for the purpose with the State being the majority shareholder and a minority holding belonging to the municipalities in the system.

  Municipal services are currently governed by Decree-Law 194/2009 of 20 August, which defines possible management models: direct, delegated or concessioned. As each of these models corresponds to forms of organisation applying to several sectors of public action, the decree-law only sets out special rules aimed at safeguarding the goals that the municipal services must pursue, especially when choosing third parties to manage municipal systems. Everything that is not provided for in this decree-law is subject to the general rules on each type of utility.

  Therefore, direct management through municipal services is still governed by the law on functioning of local authority bodies (Law 169/99 of 18 September).

  Delegation of the management of municipal systems to companies in the local business sector is governed by the corporate framework scheme (Law 53-F/2006 of 29 December), which subjects these companies to commercial law, with some specialisation arising from the need to guarantee pursuit of municipal duties, control by the municipality and financial transparency. In 2011, the framework underwent a change that had an impact on the organisation of the sectors. It was made by Law 55/2011 of 15 November, which suspended the possibility of creating new companies. The special rules of Decree-Law 194/2009 of 20 August apply to the content of the contract between the municipality or association of municipalities and the local company for provision of the delegated service and the choice of and relationship with private minority partners for these companies.

On the subject of concessioned management, Decree-Law 194/2009 complements the general rules in the Code of Public Contracts regarding the choice of concessionaire, the content of the concession contract and relations between concessionaire and grantor.

Decree-Law 90/2009 of 9 April, which sought to implement one of the management models set out in the strategic plan for the supply of water and waste water management for 2007–2013 (PEAASAR II), referred therein as integration between bulk and retail services, established the framework for partnerships between the State and local authorities for the operation and management of urban systems.

- *Legal framework on regulation*: For this regulatory component, the regulator must also ensure the existence of an appropriate legal regulation framework, which is essential to its activity, as it defines the legal framework for regulatory intervention and harmonises its procedures. It may, for example, include general provisions, corporate structure, status of the members of its bodies and personnel, economic and financial framework, powers of authority, issuance of regulations, power of sanction and final and transitory provisions.

Regarding regulation of the sectors in Portugal, the ERSAR's organic law was approved by Decree-Law 277/2009 of 2 October, which extended its intervention to all utilities of these services and reinforced an enlargement process that had been in progress for several years (Law 53-A/2006 had already set out that local companies were bound to regulation of the sectors, though without stipulating the form of intervention).

In 2010, Regional Legislative Decree 8/2010/A of 5 March set up a similar regulator for the Azores Autonomous Region, Entidade Reguladora dos Serviços de Águas e Resíduos dos Açores (ERSARA).

More recently, parliament approved Law 10/2014, with the new ERSAR bylaw. It establishes it as an independent administrative body with more organisational, functional and financial independence, reinforces its powers of regulation, especially in terms of typical legal instruments (setting tariffs and binding instructions, etc.), its regulatory competences, power to impose penalties, settlement of disputes and publication of information and consolidation of the universal nature of ERSAR's intervention on around 500 utilities, a 700% increase on the 60 or so concessions previously regulated.

- *Tariff regulations*: For this regulatory component, the regulator must also ensure the existence of appropriate tariff regulations aimed at economic and financial rationality and harmonisation of tariff structures and it must define applicable principles, tariff moderation mechanisms, rules on setting tariffs, progression of blocks, access to social and large-family tariffs, treatment of domestic and non-domestic users, objective and subjective incidence of tariffs, bases for calculation, billing of services and user relations.

In Portugal, ERSAR approved the new tariff regulations for solid waste management services applicable to all State-owned and municipality-owned utilities providing those management services, covering direct management, delegated management and concessioned management. In the preparation of tariff regulations, the solutions adopted were designed to promote gains in productive efficiency in the sectors, more rational investment decisions in articulation with national strategic goals and clearer, more transparent rules, considering the specificities of ownership of each entity and the management model adopted. The new tariff regulations for water services are currently being drafted.

- *Quality of service legislation*: The regulator must also ensure the existence of appropriate quality of service legislation for this regulatory component, defining minimum levels of quality of the services provided to users and compensation payable to them in the event of non-compliance.

- *Water quality legislation*: The regulator must also ensure the existence of appropriate legislation on the quality of water for human consumption for this regulatory component, defining the rules to be obeyed by utilities in the supply of water to users, via control of physical, chemical and bacteriological parameters. Their content may, for example, include general provisions, water quality obligations, a water quality control programme, non-compliance, test laboratories, oversight and penalties and additional, transitory and final provisions.

  In Portugal the framework on water quality for human consumption is set out in Decree-Law 306/2007 of 27 August, which lays down the main obligations of utilities of systems of water for human consumption and defines the competences of the water and waste services regulator as the competent authority for the coordination and oversight of enforcement of this decree-law, in articulation with the health authorities. The regulator also published support documents (technical reports, technical guides and recommendations) related with water quality for human consumption.

- *Commercial relations legislation*: In addition to general user protection legislation, there must be regulations on commercial relations that set out the duties and obligations of the two parties, the utility and user.

- *Technical legislation*: The regulator must ensure the existence of appropriate technical legislation for this regulatory component that sets out the rules on the design, construction and operation of infrastructure. The content on water supply systems must address the design of the systems, the basic elements for sizing distribution networks, accessory elements of the network and additional facilities, such as abstractions, treatment facilities, reservoirs and pumping facilities. The content on public waste water management systems must address the design of the systems, basic sizing elements, collector networks, accessory elements of the network, additional facilities and final disposal of storm water and domestic waste water. The content on solid waste management systems must address the design of the systems, basic sizing elements and optimisation of collection circuits.

  In terms of legislation vis-à-vis technical rules on water services, Regulatory Decree 23/95 of 23 August is still in effect although it was partially derogated by Decree-Law 194/2009.

- *Water resources legislation*: The regulator must ensure the existence of legislation on the management of water resources for this regulatory component.

  In Portugal, ownership of water resources is defined in Law 54/2005 of 15 November, which defines the public water domain, the public maritime domain, public lake and river domain and their ownership and establishes administrative easements on private parcels of beds and banks of public waters.

  The fundamental law on this matter is the Water Law approved by Law 58/2005 of 29 December, which transposes to Portuguese law Directive 2000/60/EC of the European Parliament and of the Council. It lays the foundations for sustainable water management and defines the institutional framework via harmonisation with the principle of the water basin catchment area as the main planning and management unit and setting up five river basin district administrations, which were given powers in terms of licensing and oversight of water resources. The water resources authority was given regulatory and coordinative functions as the national water authority.

  The Water Law was complemented by a series of laws, including Decree-Law 226-A/2007 of 31 May. Starting from the principle set out in the Water Law that all activities with significant impact on the state of waters can only be undertaken under a permit for use, this decree-law regulates the conditions in which an authorisation, licence or concession is granted and private uses of the water domain that also require one of these. This decree-law also set up the national information system of water resource use permits, an up-to-date inventory of existing uses managed by the water resources authority and establishing the obligation of entities with power to issue permits to register the same in this system.

The legislation on the water resource charge set out in the Water Law was established by the economic and financial framework on water resources (Decree-Law 97/2008, of 11 June).

It is important to note regarding the treatment of waste water, in transposition of an EU directive (Directive 91/271/EEC of the Council of 21 May 1991), Decree-Law 152/97 of 19 June, which defined requirements for the collection, treatment and discharge of waste water into the aquatic environment and divided water bodies into sensitive and less sensitive areas. This decree-law set out that the discharge of waste water could only be licensed, with some exceptions, after undergoing at least one secondary treatment. This made a substantial change necessary in the existing network of infrastructures at the time.

- *Waste legislation*: The regulator must ensure the existence of legislation on waste management for this regulatory component.

In Portugal, the general framework on waste, set out in Decree-Law 178/2006 of 5 September, was substantially amended by Decree-Law 73/2011 of 17 June, which republished it. The aim of these amendments was to increase prevention of waste production and encourage its reuse and recycling in order to prolong its use in the economy before returning it in appropriate conditions to the natural environment. The amendment resulted from the need to transpose Directive 2008/98/EC of the European Parliament and of the Council of 19 November 2008, which replaced Directive 2006/12/EC, in order to clarify key concepts such as the definition of waste, recovery and disposal, to reinforce measures to be taken to prevent waste, introduce an approach considering the entire lifecycle of products and materials and not only the waste phase, and also focus on the reduction of environmental impacts of waste generation and management, thereby stepping up its economic value.

The preparation of PERSU II, an instrument that reflects the review of the strategies set out in PERSU I and the national strategy for the reduction of biodegradable solid waste intended for landfills (ENRRUBDA) for 2007–2016 in mainland Portugal, was regarded as an imperative challenge to provide the sectors with clear goals and guidelines and a development strategy that might make the intervention of the different agents involved coherent, balanced and sustainable.

On the basis of the principle of the producer's extended responsibility, specific flows of waste with high recycling potential have been identified and regulated. Their management is the responsibility of those who produce or market the products that originate them. In many flows, integrated management systems financed by these producers may be created. The flow of packaging and packaging waste was the first to be regulated (Decree-Law 366-A/97 of 20 December), followed by others, such as management of used batteries and accumulators, used tyres, waste from electric and electronic equipment, used mineral-based or synthetic oils, used cooking oil, scrap vehicles and management of construction or demolition waste.

Decree-Law 183/2009 of 10 August sets out the legal framework on the disposal of waste in landfills. It steps up compliance with the principle of a waste management hierarchy and provides for minimisation of the disposal in landfills of waste with potential for recycling and recovery by restricting admission of waste included in the licence over a preset period. The target of reducing biodegradable solid waste in landfills to 35% of the total weight of biodegradable solid waste produced in 1995 was postponed to 2020. It also makes changes in the licensing procedure for landfills and its articulation with frameworks on environmental impact assessments and integrated prevention and control of pollution.

Also established in 2009 is the current framework on the use of sewage sludge on agricultural land (set out in Decree-Law 276/2009 of 2 October). Its most significant innovation was the simplification and acceleration of the activity's licensing procedure, which was now based on the sludge management plan that identifies farm holdings where it will be used, among other aspects.

The law on constitution, management and operation of the organised waste market was published in Decree-Law 210/2009 of 3 September, later amended by Decree-Law 73/2011 of 17 June. Following the framework established by the general waste management law (Decree-Law 178/2006), the organised waste market is a trading area made up of several platforms on which waste transactions recognised by the Portuguese environment agency as having the right sustainability and safety conditions are processed. It also defines financial and administrative incentives to join the organised waste market.

- *Environmental legislation*: The regulator must ensure the existence of environmental legislation for this regulatory component.

It is important to make a brief reference to the legal frameworks in Portugal on environmental impact assessment, environmental licensing, environmental offences and environmental responsibility, in terms of across-the-board environmental management instruments. All these instruments had already been set out in the Basic Law on the Environment (Law 11/87 of 7 April), the first strategic environmental law in Portugal. This required however implementing rules, which only occurred (with the exception of the framework law on environmental offences) when spurred on by EU directives.

The law on environmental impact assessments set out in Decree-Law 69/2000 of 3 May, which was reviewed and re-published in Decree-Law 197/2005 of 8 November (considering Directive 2003/35/EC), requires a prior environmental impact procedure for a substantial number of projects for infrastructures in public water and waste systems. This means greater safety in decisions about their location and technologies but also a need for longer implementation times for these projects.

Transposing the Directive on integrated pollution prevention and control (Directive 96/61/EC of the Council of 24 September), Decree-Law 173/2008 of 26 August regulates the procedure for issuing environmental licences in order to prevent or minimise emissions into the air, water and soil from certain activities, including a number of waste management operations. This licensing is articulated and requires application of other specific licensing frameworks.

The framework on environmental offences (set out in Law 50/2006 of 29 August) set up a specific framework for these offences, which until then had been governed by the general framework on administrative offences (Decree-Law 433/82 of 1 November). This one is different, among other aspects because it allows the accountability of owners, directors or managers of legal persons and because of a substantially higher limit on penalties for very serious environmental offences.

There is also the framework on liability for environmental damage approved by Decree-Law 147/2008 of 29 July, which transposed to Portuguese law Directive 2004/35/EC of the European Parliament and of the Council of 21 October, which approved, on the basis of the polluter pays principle, the framework on environmental responsibility for the prevention and reparation of environmental damage. Activities likely to generate environmental responsibility include operation of facilities subject to environmental licensing, waste management operations and the discharge of pollutants into surface or underground waters. Utilities engaging in these activities were obliged as of 1 January 2010 to set up financial guarantees allowing them to accept environmental responsibility for their activity.

- *Consumer protection legislation*: The regulator must ensure the existence of user protection legislation for this regulatory component that defines mechanisms for protecting users of these services that take the form of legal obligations especially applicable to their providers. Its content can, for example, include available information, prohibition of charging prices that do not match costs, obligation to give prior notice of suspension due to non-payment, frequency of billing, minimum payment time and especially short times for expiry of debts.

In Portugal, Law 23/96 of 26 July, also called the Essential Public Services Law, provides for mechanisms to protect essential public service users, in which a user is considered a natural or legal person to whom the service provider is obliged to provide the service. This is a broader concept than that of 'consumer' set out in the also important Consumer Protection Law (Law 24/96 of 31 July), which only include persons to whom goods are supplied, services are provided or any rights are transferred for non-professional use by a person who works in an economic activity for gain.

In the initial version of Law 23/96, the list of essential public services only included the supply of water, electricity and telephone. It was only in 2008, with the amendment made by Law 12/2008 of 26 February, that the waste water management and solid waste management services also fell under this framework.

The third amendment to the Essential Public Services Law made by Law 6/2011 of 10 March established the necessary arbitration of disputes over essential public services when requested by users. Law 44/2011 of 22 June also amended the Essential Public Services Law in 2011, though it had no great impact on the water and waste sectors.

Regarding users' right to complain about services, an important feature was the complaint book introduced by Decree-Law 156/2005 of 15 September, which made it compulsory (the so-called Red Book) for all suppliers of goods and services involving contact with the public, with the exception of the Public Administration, which was only covered by this law after publication of Decree-Law 317/2007 of 6 November. This decree-law also extended the time for sending the original complaint to the regulator from five to ten days, thus allowing the service provider to attach proof of the response to the user to the original.

In 2009, Ministerial Order 866/2009 of 13 August set up the common telematic information network (RTIC), an online portal for regulators to register complaints received and their analysis and treatment procedure. This platform replaces the regulators' obligation to periodically report information to the Directorate-General for Consumers on complaints received. It allows that entity to view complaints in real time and obtain statistics on complaints. It also allows complainants and entities subject to a complaint to view the status and progress of complaints online.

Returning to Decree-Law 194/2009 of 20 August, it includes a chapter exclusively devoted to relations between utilities for water and waste services and end users. It deals with issues such as the obligation to provide services and connect to public networks, contracts, admissibility and conditions of suspension of service, metering and complaints.

Finally, Ministerial Order 34/2011 of 13 January established the minimum content of service regulations on the provision of water and waste services to end users. Approval of this order was expressly provided for in Article 62 (1) of Decree-Law 194/2009, which indicates that the rules on the provision of the service to users must be set out in service regulations to be approved by the incumbent entity.

- *Competition legislation*: The regulator must ensure the existence of legislation on competition for this regulatory component.

In Portugal, an important milestone in the development of the competition sector was the creation of the Competition Authority by Decree-Law 10/2003 of 18 January, which had an across-the-board mission to defend competition. The authority's jurisdiction covers all sectors of economic activity and there are mechanisms for articulation with sectoral regulators when necessary.

Following the creation of this institutional framework, a new competition law was approved by Law 18/2003 of 11 June, which regulates the conditions and procedures of the Competition Authority with regard to oversight of company agreements, abuse of dominant positions, State aid and prior appreciation of company concentration operations. Reflecting EU law on competition, this law considers State-owned companies and companies to which the State has granted special or exclusive rights are bound to comply with competition law, to the extent that the enforcement of these rules is not an obstacle to fulfilment, by law or de facto, of their particular mission.

Finally, the legislation on public contracts, which has a broader object and goal, is also an instrument aimed at guaranteeing competition in the public contract market. The Public Contract Code was approved by Decree-Law 18/2008 of 29 January and came into force in June 2008. It has been amended several times and has been regulated (through Ordinance 959/2009 of 21 August).

- *Regulatory procedures legislation*: For this regulatory component, the regulator must ensure the existence of appropriate legislation on regulatory procedures, which must define in detail the procedures and schedules inherent in the regulator's relations with the utilities that it regulates in the pursuit of the powers and competences granted to the regulator by law, in legal and regulatory terms.
- *Recommendations*: The regulator must also draft voluntary recommendations on specific matters to complement legislation and regulations, clarify pending issues and specify procedures to assist the utilities and other agents of the sectors.

In Portugal, ERSAR has published recommendations in recent years, which are available on its website (www.ersar.pt) mostly in Portuguese:

- *Recommendation 1/2005: Prevention of possible negative effects of drought on quality of water distributed.*
- *Recommendation 2/2005: Control of lead in water for human consumption.*
- *Recommendation 3/2005: Control of iron and manganese in water for human consumption.*
- *Recommendation 4/2005: Control of arsenic in water for human consumption.*
- *Recommendation 5/2005: Alternative method for analysing coliform bacteria and Escherichia coli.*
- *Recommendation 6/2005: Procedure for utility when water quality parameters are not met.*
- *Recommendation 7/2005: Control of bromides in water for human consumption.*
- *Recommendation 8/2005: Procedure for sampling water for human consumption in public supply systems.*
- *Recommendation 1/2006: Selection of engineering planning services in the water and waste sectors.*
- *Recommendation 2/2006: Good practices in the acquisition of products used for treatment of drinking water.*
- *Recommendation 1/2007: Management of septic tanks in private solutions for the disposal of waste water.*
- *Recommendation 2/2007: Use of treated waste water.*
- *Recommendation 3/2007: Accounting and contract procedures.*
- *Recommendation 4/2007: Billing of bulk waste water treatment services in systems with storm water inflows.*
- *Recommendation 5/2007: Disinfection of water for human consumption.*
- *Recommendation 1/2008: Communication and correction of non-compliance with quality parameters of water for human consumption.*
- *Recommendation 2/2008: Correction of aggressiveness of water for human consumption in small population clusters.*
- *Recommendation 3/2008: Control of quality of water for human consumption in private supply systems.*
- *Recommendation 1/2009: Formation of tariffs for end users of drinking water supply, waste water management and solid waste management ('tariff recommendation', available also in English).*
- *Recommendation 1/2010: Contents of invoices for drinking water supply, waste water management and solid waste management services provided to end users ('invoice contents').*
- *Recommendation 2/2010: Criteria for calculating tariffs for end users of drinking water supply, waste water management and solid waste management ('calculation criteria').*
- *Recommendation 3/2010: Procedure for collecting samples of water for human consumption in supply systems.*
- *Recommendation 1/2011: Alternative colilertr-18/quanti-trayr method for finding and quantifying coliform bacteria and Escherichia coli in water for human consumption.*
- *Recommendation 2/2011: Technical specification for certification of water for human consumption.*
- *Recommendation 3/2011: Quarterly publication of quality data of water for human consumption.*
- *Recommendation 4/2011: Assessment of risk when determining taste in samples of water for human consumption.*

- *Annual report*: The regulator must draft an up-to-date annual report on water and waste services and the rules of the sectors in terms of legislation, regulations and recommendations for everyone needing reliable information in order to define policies or business strategies to assess the services actually provided to society.

  In Portugal, ERSAR has published an annual report on water and waste services in Portugal (RASARP) since 2004. It includes a section that identifies legislation, regulation and recommendations and is available on its website (www.ersar.pt).

The regulatory activity described, undertaken in compliance with the regulatory procedures mentioned and with recourse to these regulatory instruments, enables the regulator to effectively and efficiently achieve its goals of contributing to clarifying the rules of the sectors, one of the components of the regulation model for the public services for drinking water supply, waste water management and solid waste management.

7.5 REGULATORY SYNERGIES

Regulatory contribution to legislation of the sectors articulates closely with other components of the regulation model. It is structural and so it impacts on all of them and the decisions made in the contribution to better organisation of the sectors and information gathered in the regulation of legal and contractual compliance, economic regulation, quality of service regulation, quality of water for human consumption regulation and interaction with users regulation can lead to intervention decisions because of a possible need to amend legislation, regulations or recommendations.

7.6 SUMMARY

This chapter gave a detailed description of one of the components of the proposed regulatory approach in the framework of structural regulation, the regulatory contribution to legislation of the sectors, including its objectives, activities, procedures, instruments and synergies.

The next chapter describes another of the components of the proposed regulatory approach, also in the framework of structural regulation, the regulatory contribution to information on the sectors.

Chapter 8

Regulatory contribution to the information of the sectors

8.1 INTRODUCTORY NOTE

As part of this integrated approach (model RITA-ERSAR), this chapter describes in more detail the regulatory contribution to information on the sectors, one of the components of the regulation model for the public services of drinking water supply, waste water management and solid waste management.

8.2 REGULATORY GOALS

The aim of this component of the regulatory model must be to promote the provision of information on the sectors by regularly drafting and disseminating accurate information that is accessible to all concerned. It is the regulator's responsibility to consolidate a culture of concise, credible, accessible and easily interpreted information for all agents, such as service users and utilities, regardless of the forms of management used to provide the services.

This component of the regulatory model thus helps to achieve the public policy goals, defined in 3.3, namely provision of information, and also fulfil public service obligations, defined in 2.2, such as adoption of rules on good practices in access to information.

8.3 REGULATORY ACTIVITIES AND PROCEDURES

The main activities of this regulatory component include collection of information on the sectors and its utilities and provision of concise, credible, accessible and easily interpreted information to assist in defining public policies or strategies, improving management of services and making improvements in the sectors or promoting greater and better public participation:

- *Definition of an information system*: The regulator must define the information that is relevant to the sectors and, based on it, ensure the existence of an information system for collecting this information effectively, efficiently and homogeneously on all utilities. This means that clear rules on preparation and regular reporting of information must be defined so that utilities can define and implement an appropriate procedures.

   The success of an information system depends largely on the regulator's ability to make it clear why it is requesting the information and then proactively send the recipients the results generated with this information, to demonstrate de added value. In fact, it is important for a utility to be aware that the information it reports has a rational purpose and that the results will be published, which is a supplementary guarantee of the accuracy of the information reported.

In addition to the information requested by the regulator needing to be only that which is strictly necessary, it is important to prevent utilities from sending too much information, which may, intentionally or not, prejudice the effectiveness and efficiency of regulatory intervention.

- *Obtaining, validating, processing and publishing information*: Information should be obtained in interactions between utilities and the regulator as part of the components of legal and contractual and economic regulation, regulation of quality of service, quality of water for human consumption and interaction with users. It is up to the regulator to collect, audit, process and publish the information. Audits consist of a careful and systematic analysis of the information regularly or sporadically provided by utilities, in order to check for reliability. The data may cover different areas, for example legal, financial and quality.

  Finally, the regulator must disseminate the information for professional use, in an annual report on water and waste services, for example, in a way that fills in gaps in information, including a general characterisation of the sectors and of its utilities in legal, contractual, financial and economic aspects and in terms of quality of service, quality of water for human consumption and interaction with users.

  It must also provide information to all users of these services giving them a clear idea of the utility that provides the service, the quality of the water supplied to them, the prices charged, quality of service and complaints processed. This information should also allow them to quantify this reality clearly and also to compare the performance of the providing entity with that of others. This information shows users whether the entity providing them the service is efficient and, if not, enable them to demand improvements.

  The regulator must also disseminate information via events in the sectors, such as lectures and papers, and in the media, such as interviews, answers to questions and press releases.
- *Making improvements in the sectors*: The regulator must, based on experience acquired in collecting information on the sectors and its utilities, make periodic reassessments, identify the need for and, if necessary, implement any changes to the content of the information or procedures for obtaining, validating, processing and publishing information in order to improve the functioning of the sectors and the activity of the utilities.

The regulator must follow a specific, clear, rational procedure for its regulatory contribution to information on the sectors set out in the rules on regulatory procedures.

Regulation of information on the sectors must apply to each utility over time periods based on the different cycles of behavioural regulation of utilities, in terms of legal and contractual and economic regulation, regulation of quality of service, quality of water for human consumption and interaction with users, without prejudice to casuistic, needs-based interventions.

There follows a possible example of a regulatory procedure that can be followed by the regulator. It is divided into stages:

- The regulator must ensure the existence of an information system and clear rules on the collection, validation, processing and dissemination of information, in the form of legislation and additional documentation, always taking into account the need to streamline the system and the existence of a single location for reporting data to facilitate reporting by utilities.
- The regulator must collect the different flows of data from the different cycles in the regulatory model by using the appropriate procedures and modules for the information system.

- If necessary, the regulator must complement collection of data from the different components of the regulatory model with any relevant additional data.
- If justified, the regulator must levy penalties against a utility that fails to send information, as set out in legislation, or that send incorrect information.
- The regulator must periodically disseminate the results available in its information system for professional use, for example in its annual report on water and waste services, circulate results for non-professional use, mainly for users, on its website (www.ersar.pt), and send information to the national statistical authority.
- The regulator should, based on its experience, periodically identify and promote implementation of any amendments to legislation, regulations or others defined in the regulatory activity to improve information on the sectors.

Figure 8.1 shows the different stages in the procedure for regularly preparing and publishing information.

Figure 8.1 Regulatory contribution to the information of the sectors.

8.4 REGULATORY INSTRUMENTS

The regulator must promote, develop and use the most appropriate instruments for regularly preparing and publishing information. There follow some examples of regulatory instruments:

- *Information system*: In view of the large amount of information that this activity generates, regulatory effectiveness and efficiency must be improved by using an information system with external modules for legal and contractual regulation, economic regulation, regulation of quality of service, quality of water for human consumption and interface with users. It must include also internal regulation modules for management of documents, processes, administration, finances and filing used by the organisation for its own management. Furthermore, the regulator must have a website for publishing information independently and credibly to all concerned, which for example allows that they can compare this information between utilities. This chapter describes the information system in more detail in view of their special importance to regulation.

  ERSAR has a powerful information system that acts as an interface between the regulator, utilities and other administrative entities in Portugal and manages around a million pieces of data every year.

- *Annual report on water and waste services*: The regulator must issue reliable periodical reports on the situation and development in the sectors, which are regularly renewed, for all those who require reliable information to help define policies or business strategies or to assess services actually provided to society. One way of doing so is to publish an annual report on water and waste services essentially for professional use, capable of filling information gaps. It should include a general characterisation of the sectors and utilities in legal, contractual, financial and economic terms and in terms of quality of service, quality of water for human consumption and interaction with users.

  ERSAR has published an annual report on water and waste services in Portugal (RASARP) since 2004. It includes five sections with comprehensive information on the sectors situation and is available on its website (www.ersar.pt).

- *Information for non-professional use*: Every year, the regulator must publish up-to-date information for non-professional use on its website. This data must be presented in an intuitive, comparable form for users. It should also use awareness leaflets in simple and media, with non-technical language to communicate with society and other agents for users, thereby contributing to the transparency of the sectors.

  In Portugal, the ERSAR website (www.ersar.pt) has permanent interactive apps offering users easy access to data on user tariffs and costs, quality of service, quality of water for human consumption and complaints. One application for mobile phones is also available.

- *Information through the media*: The regulator must communicate with society and other agents in the media by granting interviews, issuing news and press releases and answering the numerous queries, thereby contributing to the up-to-date information on the sectors.

  In Portugal, ERSAR has published articles, papers, interviews and press releases in the 'ERSAR Intervenções Públicas' series every year since 2004. The annual editions from 1999 are available on its website (www.ersar.pt).

- *Information via events in the sectors*: The regulator must communicate with society and other agents via articles in publications and papers at technical events to disseminate up-to-date information on the sectors, thereby contributing to the transparency of the sectors. This communication is very important and warrants great attention from the regulator.

In recent years in Portugal, ERSAR has published public reports in the 'ERSAR: Intervenções Públicas' series. The annual editions from 1999 are available on its website (www.ersar.pt).

- *Rules on regulatory procedures*: The regulator must have rules on regulatory procedures that define in detail the procedures and schedules concerning relations with the utilities under its regulation as part of the duties and competences invested in it by law, particularly in terms of information.
- *Penalties system*: There must be a penalties system allowing the regulator to impose penalties on utilities for acts or omissions infringing laws on the provision of information, including failure to provide it or provision of false, inaccurate or incomplete information that is requested by the regulator or that which is required by law.

The regulatory activity described, undertaken with the above-mentioned regulatory procedures and these regulatory instruments, allows the regulator to effectively and efficiently achieve the goals of its contribution to information on the sectors, one of the components of the regulation model for public drinking water supply, waste water management of solid waste management services.

8.5 REGULATORY INFORMATION SYSTEM

8.5.1 Overview

There is always asymmetry in the access to information between regulated entities and the regulator, which needs to be minimised and prevented since regulation requires abundant and accurate information for it to be effective.

The large amount and especially the diversity of information pose additional challenges in collecting and systematising it. This is due to the different regulation components for which the regulator is responsible, such as legal and contractual and economic regulation, regulation of quality of service, quality of water for human consumption and interaction with users, and also owing to the frequently varied types and characteristics of utilities.

It is therefore essential for the regulator to structure and implement an information system that enables it to perform its duties properly. The generic goals are:

- Identify the most relevant information for the sectors and for regulation.
- Facilitate the collection, validation and processing of information to support decision making by standardising and optimising the regulator's internal procedures so as to improve the quality and efficiency of regulation and thereby reduce its operating costs.
- Provide a platform for effective relations between the regulator and the utilities by allowing them access to a single location for reporting data and viewing information from the regulator.
- Facilitate access to information for all other agents of the sectors.

Due to the need for articulation between the different components of the regulator's work, the information system must be able to integrate information between the different modules and allow cross validation of information provided by utilities for different purposes.

The information system must also be a repository for information that feeds other interfaces, such as the website and the regulator's geographical information system.

It is not advisable to have several information systems regarding those services in the same country managed by different entities with different perspectives, overloading the utilities with multiple requests for information that are sometimes the same, other times contradictory and confusing potential users with apparently identical but actually different information with varying degrees of reliability. The regulator must have special responsibility in creating this single information system based on a single location for reporting data.

The information gathered in the system must also be usable as an official source of statistics by the national statistical authority in the area of water and waste services, which reduces the overall costs of collecting these data at national level.

The regulatory information system must provide stability, speedy implementation, modular implementation, scalability, security, advanced access management and streamlining of costs.

As shown in Figure 8.2, it may consist of:

- External modules of the information system covering legal and contractual and economic regulation, regulation of quality of service, quality of water for human consumption and interaction with users.
- Internal modules of the information system covering document management, process management, administrative management, financial management, control of regulatory activity and archive management used by the organisation for its own management.
- A website accessible to the general public that provides the most relevant information on the water and waste sectors.

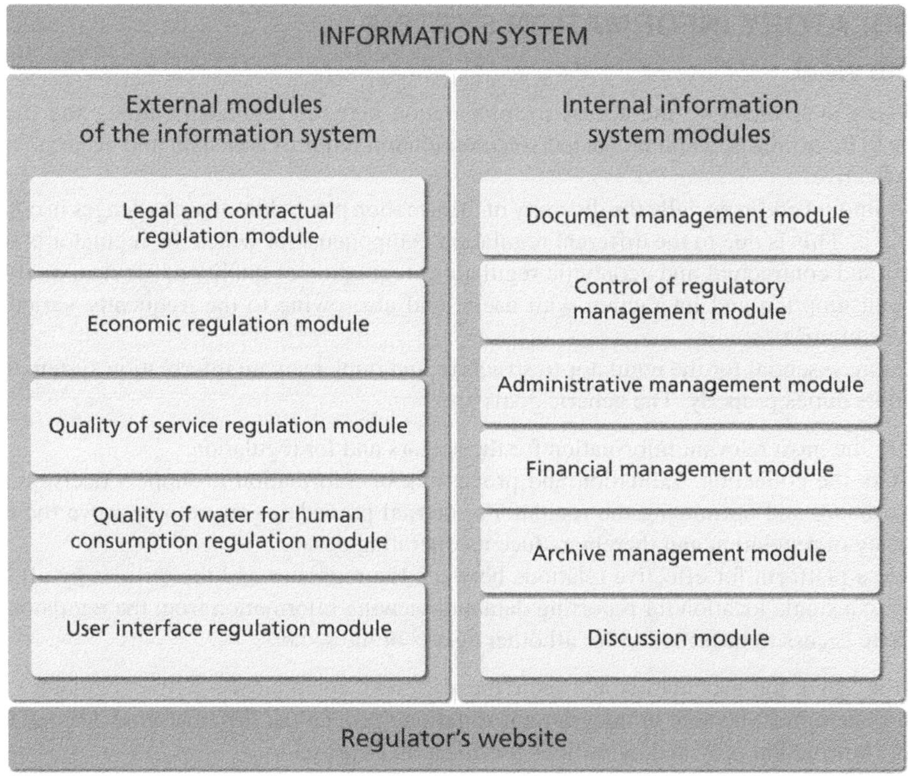

Figure 8.2 Regulatory information system.

8.5.2 External modules of the information system

There follows a description of the modules for legal, contractual and economic regulation, regulation of quality of service, quality of water for human consumption and interaction with users.

Legal and contractual regulation module

This module must be a repository of legal and contractual information on the utilities with which the regulator interacts.

Each utility must be described with the title that qualifies it to provide services, the contractual documents by which it is bound, the services it provides, the geographical areas it serves, the population it covers and all its contact details.

This module is fed by the legal and contractual regulation activity, namely by the utilities that have to send periodic information and by the regulator, to the extent that it monitors the whole lifecycle of the utilities, which covers the design, possible tender procedure, the contract, management of the service and amendments to or termination of the contract.

In addition to being naturally accessible to the regulator and utilities, part of this information is also integrated into the regulator's website to make it available to users and other agents of the sectors.

Economic regulation module

This module must be a repository of economic and financial information on utilities and support documentation for the regulator's regulation process.

Each utility must be characterised with the past and present tariffs in effect in its geographical area, documentation and data on accounts and indicators that reflect its economic and financial situation and sustainability.

This module may also be very important to utilities, especially smaller ones with fewer resources, as a way of structuring the collection of information and a management tool within the entities themselves. It can improve their knowledge of costs and earnings from providing the service and consequently identify aspects requiring higher efficiency and ensure the economic sustainability of the service.

Furthermore, it should be possible to export reports so that the information can be processed and published in the annual report on water and waste services and in studies conducted by the regulator.

This module is fed by the economic regulation activity itself, namely by the utilities that have to send information periodically and by the regulator in that it monitors the definition of tariffs and the utilities' actual accounts.

In addition to being naturally accessible to the regulator and utilities, part of this information is also integrated into the regulator's website to make it available to users and other agents of the sectors.

Quality of service regulation module

This module must be a repository of quality of service information provided to the utilities' users, which reflects social sustainability (suitability of the user interface), sustainability of management of the service and environmental sustainability, and support documentation for the regulator's regulation process.

Each utility must be characterised with the data necessary for assessment, the utility's profile, the system's profile, context factors, quality of service indicators and reference values.

This module may also be very important to the utilities, especially smaller ones with fewer resources, as a way of structuring collection of information and a management tool within the entities themselves. It can improve internal processes consequently identify aspects requiring higher efficiency and consequently improve their efficiency, which naturally results in better quality of service.

Furthermore, it should be possible to export reports with an assessment of the indicators compared to the reference intervals so that the information can be processed and published in the annual report on water and waste services and in studies conducted by the regulator.

This module is fed by quality of service regulation activity itself, namely by the utilities that have to send information periodically and by the regulator in the monitoring of their performance.

In addition to being naturally accessible to the regulator and utilities, part of this information is also integrated into the regulator's website to make it available to users and other agents of the sectors.

Quality of water for human consumption regulation module

This regulation module must be a repository of information on the quality of water for human consumption provided to the utilities' users and support documentation for the regulator's regulation process.

Each utility must be characterised with the data necessary for assessment, reference values and non-conformities. This module must allow real-time reporting of any non-conformities that endanger public health in order to enable responsible entities to find out almost immediately about these situations.

This module may also be very important to the utilities, especially smaller ones with fewer resources, as a way of structuring the collection of information and taking immediate action in the event of a potential risk to public health and a management tool within the entities themselves. It can improve internal processes and consequently improve their efficiency, which naturally results in better quality of service.

Furthermore, it should be possible to export reports with an assessment of the indicators compared to the established limits so that the information can be processed and published in the annual report on water and waste services and in studies conducted by the regulator.

This module is fed by the regulation of quality of water for human consumption itself, namely by the utilities that have to send information periodically and by the regulator in the monitoring of the performance.

In addition to being naturally accessible to the regulator and utilities, part of this information is also integrated into the regulator's website to make it available to users and other agents of the sectors.

User interface regulation module

This module must be a repository of information on the utilities' user interface and support documentation for the regulator's regulation process.

Each utility must be characterised with the data necessary for assessing compliance with its obligations to users. This module is for management of complaints from users of water and waste services and allows the entirely computerised exchange of information and subsequent online indication of the current status of the complaints to users.

This module should allow the cutting of red tape in relation to the circulation of information between utilities and the regulator regarding the management of complaint processes, thereby speeding up their handling by the regulator and utilities, with obvious benefits for the complainants. This tool should also allow complainants to monitor their complaints online, which saves on internal resources needed to answer complainant's requests. The module must also manage the internal circuit for analysis and issue of the decision on complaints by the regulator, along with the extraction of statistics on the number of complaints per utility, subject, stages of the process, response times, among others. This should be under constant analysis by the regulator for negative aspects related with the utility in order to correct problems reported by users. This information is also used for the internal assessment of response times by the regulator and its employees.

This module may also be very important to utilities, especially smaller ones with fewer resources, as a way of structuring complaint management processes and a management tool within the entities themselves. It can improve the utilities' management processes and consequently improve their efficiency, which naturally results in better quality of service.

Furthermore, it should be possible to export reports with an assessment of the handling of complaints so that the information can be processed and published in the annual report on water and waste services.

This module is fed by the regulation of the user interface itself, namely by the utilities that have to send information periodically and by the regulator, in that it monitors their performance.

In addition to being naturally accessible to the regulator and utilities, part of this information is also integrated into the regulator's website to make it available to users and other agents of the sectors.

8.5.3 Internal information system modules

There follows a description of the module for document management, control of the management of regulatory activity, administrative management, financial management, archive management and discussion management.

Document management module

This module should handle the regulator's document management ensuring the register and circulate documents between the regulator's staffs and optimise the circulation of information, thereby cutting red tape in the internal circulation of documentation with substantial savings of time and resources. All the circulation of documents received by the regulator and all documentation produced and sent by the regulator must be based on this module. Considering the vast number of attributes characterising them and the connection between documents, searching for a given document is thus quite fast.

It should also aggregate related documents, associate specific information from a certain process and monitor time limits in the different phases of processes. This module must also be integrated with other operating modules for collecting external information, for example receipt and management of complaints from water and waste service users. The application is used, for example, in the management of complaints made by water and waste service users to speed up internal procedures.

This module should also aggregate related documents so that the relationship between these documents and the flow of information associated with a particular matter can be easily identified. All those concerned in the process will be automatically informed of the status of the matter and be able to answer queries from third parties.

Specific types of process must therefore be identified along with each information flow. Specific time limits must be set for each phase of the process.

This means that information is centralised and available online to all concerned, thereby facilitating the incremental work done by staffs in the different units.

Control of regulatory management module

This module should allow control of regulatory management by the regulator's staffs, by monitoring the work flow and response times and comparing the volume of work of different departments and staffs, among others. It should be able to assess each employee's level of performance in response to requests allocated to them, namely how many requests are solved within the deadline, and aggregate this information at unit level and then at organisation level.

Administrative management module

This module should handle the regulator's administrative management, for example employee attendance and work hours, in order to provide information for wages processing. It should also manage employees requests for leave and record absences on leave, manage requests for assistance and other requests from internal and external portal users and internal requests for support for information technologies, office supplies, booking meeting rooms, human resources and others.

Financial management module

This module should handle the regulator's financial and budgetary management, including analytical accounting and management of the collection of regulation charges from utilities, and record information on the levels of activity on which the charges are based.

Archive management module

This module must be used to file technical, economic and legal documentation, images, news articles and diverse information on the sectors and the regulator's activity. For example, it should be possible to search for relevant documentation and all legislation related to the regulator's activity and the water and waste services sectors. The information provided must be integrated with the regulator's website and automatically made available to the public.

Discussion module

This module should be used for forums for clarification and the discussion of a variety of matters for the exchange of information with internal users, analysis of questions regarding the organisation's functioning or external users categorised by regulatory cycle regarding new legislation, for example. It should also be used for surveys of internal and external users and subsequent production of statistics. This is an important tool for assessing users' satisfaction with the information system.

8.5.4 Regulator's website

The regulator's information system must also include its website, which is accessible to the general public. It should be used to disseminate the most important information on its activity and the water and waste sectors. It is one of the modules that receive processed information on the regulator's regulatory cycles. The website must be fully automated for receiving information uploaded into the information system. This integration will speed up the uploading of contents onto the site and ensure that the information is always up to date.

The website should favour provision of information to the general public, particularly service users and should therefore be user friendly with accessible, clear, concise information. This information must also allow benchmarking of information between utilities in order to create healthy pressure that will improve the efficiency and effectiveness of services.

It is advisable, in order to make technical information available to other people involved in the sectors, to create specific areas with more detailed, workable information to allow research centres and universities to gather information for studies and processing. This has obvious advantages in terms of developing the study of subjects complementary to regulation but that can have a great impact on improving water and waste services.

It is also worth considering creating a dedicated area for the media for the provision of general information and press releases for use by the media to disseminate information.

8.5.5 Advantages of the regulatory information system

The regulator should use the information system in its internal processes, and to exchange information with utilities and other partner entities as an essential tool in its operation, because it simplifies data reporting and plays a role in the centralisation of information accessible to the regulator and society as a whole.

The added productivity that the information system can allow in the regulator's activity is one of the factors that enable it to cover a vast, diversified number of matters in the regulation of utilities, while also improving quality regulation.

The information system has a number of important advantages for the regulator, such as:

- Cutting paper use and consequently ensuring faster, more effective management of all communication.
- Elimination of the physical management of the different regulation cycles, which results not only in faster management of processes but also lower costs of physical storage and the circulation of information.
- Better access to data, which are available in the web app and therefore permanently available.
- Greater productivity in searching for information as it is no longer necessary to search physical files and it is easier to access essential information for the regulation of services and for making management applications available.
- More sharing of internal information due to the centralisation and circulation of information in the same digital format, which allows greater interdisciplinarity in operation with a clear influence on the quality of decisions.
- Shorter response times between regulator and utilities and a lower risk of non-compliance with legal obligations by either party.
- Lower costs of exchanging information with utilities and other entities with which the regulator works.
- Greater transparency in procedures that, thanks to the harmonisation of formats in which the information is received, means that different data can be processed uniformly.
- Lower probability of loss of documents in the internal circulation of information.
- Better organisation of information, which is centralised and organised in the regulator's information system.
- Greater capacity for monitoring tariffs charged by utilities, which no longer send the information at their discretion in several different formats.
- Greater capacity for monitoring the economic sustainability of utilities, which now report accounting information that is homogeneous and comparable.
- Greater capacity for monitoring and assessment of quality of service to users and resulting in greater efficiency of services.
- Improvement in the management of data on quality of water for human consumption, with greater process management capacity.
- Better management of complaints and greater response capacity to complaints received.
- Greater capacity of the regulator for validating the information reported, which may be validated in advance by the information system before being submitted to the regulator, and which can also be validated after sending by cross-referencing data from the same utility over years and between different utilities.
- Greater capacity of the regulator for processing and producing information for publication, as it can be exported automatically to more easily workable formats.
- Better capacity for production of the regulator's reports for annual circulation to the public.
- Better capacity for annual monitoring of the sectors' development in relation to strategic plans.
- Higher motivation and productivity of regulator's specialists thanks to faster handling of processes, resulting in growing involvement and commitment throughout the organisation.

The information system can be an important productivity tool for the utilities as it enables them to optimise internal procedures and provide harmonised information to the regulator, which will certainly help improve the utilities' knowledge of the services that they provide. The information system has a number of important advantages, such as:

- Possibility of using a technological tool for exchanging information with the regulator.
- Greater facility in filling in data, thanks to standardised formats for reporting information and the resulting greater capacity for compliance with legal deadlines for submitting information to the regulator.
- Equality in the processing of data reported by the utilities, thanks to the standardisation of information between entities.
- Possibility of access to data submitted and to the status of processes under way at the regulator, because the information is permanently available.
- Reduction in response times between the regulator and utilities, reducing bureaucracy and increasing efficiency in procedures along with the time necessary and associated costs for utilities when sending information.
- Greater capacity for the collection and reporting of information on costs and earnings pertaining to the provision of the service, which over time results in better definition of tariffs appropriate to the services provided.
- Greater capacity for the collection and reporting of information on user quality of service, which over time results in the utilities' greater efficiency.
- Greater ease in management of control programmes for the quality of water for human consumption, meaning that the water quality module can be used as a management instrument by the utilities in their scheduling and collection of water quality tests and also in speedier reporting of non-compliance with water quality placing public health at risk.
- Greater capacity for handling complaints by utilities, as they can benefit from the information system developed by the regulator.
- Modernisation and capacity building of utilities, in that the information system can be an important tool in boosting innovation and modernisation of utilities' procedures.
- Possibility of concrete requests for intervention by the regulator and for submitting queries and suggestions via the discussion and clarification forums.

The information system has important advantages for users of this service, such as:

- Greater facility in the publication of plentiful, consolidated information on the sectors in a form that is intuitive, up to date, clear and accessible to the public, including lists of important entities, legislation, maps of the sectors, interactive apps and documentation.
- Better perception by users of the services provided to them, for example about tariffs, quality of service or water quality.
- Faster response to risks to public health.
- More transparent information.
- Greater accountability of utilities.

For users, the information provided by the regulator means a greater balance of power with the utilities, which have much more detailed information about the service. Therefore, with the provision of information on the regulator's website, the regulator fulfils one of its main missions, which is the protection of users' interests.

Where other entities from public administration are concerned, such as environmental, water resource, public health and statistical authorities, the information system offers important advantages, such as speedy provision of the information they need for administrative processes and the provision official statistics.

Where the country and society as a whole are concerned, the information system has important advantages, such as:

- Greater probability of achieving goals set out in the sectors' strategic plans.
- Reduction in risks of public health problems thanks to greater capacity for controlling water quality.
- Lower risks of environmental problems thanks to greater capacity for monitoring discharges and pollutant emissions.
- Greater pressure on all involved to work towards better efficiency.
- Lower risk of regression in matters as sensitive as water quality, quality of service and economic accessibility of services.

8.6 REGULATORY SYNERGIES

The information on the sectors articulates closely with other components of the regulation model, as the information gathered comes essentially from legal, contractual and economic regulation, regulation of quality of service, quality of water for human consumption and interaction with users. Conversely, all these components naturally benefit from the information system in terms of regulatory analysis.

8.7 SUMMARY

This chapter gave a detailed description of one of the components of the proposed regulatory approach in the framework of structural regulation called regulatory contribution of the information of the sectors, including its objectives, activities, procedures, instruments and synergies.

The next chapter describes another of the components of the proposed regulatory approach, also in the framework of structural regulation, the regulatory contribution to capacity building in the sectors.

Chapter 9

Regulatory contribution to the capacity building of the sectors

9.1 INTRODUCTORY NOTE

As part of this integrated approach (model RITA-ERSAR), this chapter describes in more detail the regulatory contribution to capacity building of the sectors, one of the components of the regulation model for the public services of drinking water supply, waste water management and solid waste management.

9.2 REGULATORY GOALS

This component of the regulatory model must be aimed at supporting capacity building in the sectors, promoting research and development, creating endogenous innovation and knowledge and empowering human resources with appropriate vocational training for their work. Regulation should foster the country's growing scientific and technic independence in areas associated with water and waste services and even increase the value added of the sectors, possibly leading to the export of knowledge acquired and products.

This regulatory procedure is important not only as it empowers the sectors, but also conveys a more credible image of the regulator's mission and technical competence, thereby facilitating its acceptance by utilities.

This component of the regulatory model thus helps to achieve public policy goals, defined in 3.3, such as capacity building of human resources and promotion of research and development.

9.3 REGULATORY ACTIVITIES AND PROCEDURES

The main activities of this regulatory component include setting capacity building priorities in the sectors, implementing the corresponding activities and making improvements in the sectors.

The regulator must set capacity building priorities in the context of the country's development based on its special knowledge of the sectors, in terms of priority research and development, the need for studies about the sectors, the need for technical publications and the need for training of human resources.

The regulator should promote research and development in the country to foster innovation in the sectors, in partnership with universities and other national and international knowledge centres in articulation with innovation agencies. In addition, the regulator must improve knowledge of the sectors by conducting studies.

The regulator must also provide technical support to utilities by promoting training and support for third parties' events on priority, relevant issues for the sectors. It should also provide technical support to utilities by means of timely, reasoned answers to their queries. In the capacity building of specialists, particular attention should be devoted to adapting utilities to new regulatory requirements in order to ensure the quality of management practices and necessary information for regulatory purpose.

Finally, the regulator must promote technical publications, if necessary in partnership with knowledge centres, for example by publishing technical manuals and technical courses.

Provision of the results of these activities, such as studies and research and development, and attendance of training courses and access to technical publications must be free of charge, as it should be supported by revenue from regulation of the utilities. This will mean that the sectors and utilities in particular can reap greater access and benefits from them.

The regulator should, based on its experience, periodically assess, identify the need and, if necessary, promote the implementation of changes to its strategy for supporting innovation and technical capacity building, thereby playing an active, permanently up-to-date role in the sectors in terms of its regulatory contribution to innovation and capacity building.

The regulator should follow a clear, rational procedure for supporting innovation and technical capacity building in the sectors, as set out in regulations on regulatory procedures, which will benefit the entire sectors.

These are regulatory activities that are not necessarily scheduled. They are casuistic on the basis of needs that arise, though they may be included in the regulator's annual programme of activities.

Figure 9.1 shows the different stages in the procedure for supporting capacity building in the sectors.

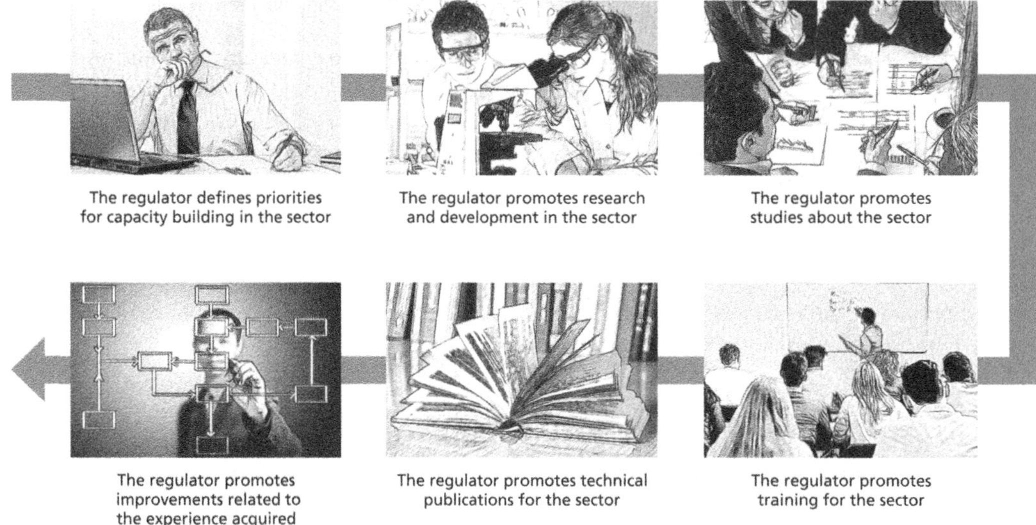

Figure 9.1 Regulatory contribution to the capacity building of the sectors.

9.4 REGULATORY INSTRUMENTS

The regulator must promote, develop and use the most appropriate instruments to support capacity building in the sectors. There follow some examples of regulatory instruments belonging to the regulator and external to the sectors:

- *Research and development projects*: The regulator must promote research and development to boost innovation in the sectors in partnership with universities and other national and international knowledge centres in articulation with national innovation agencies on the basis of the sectors' priorities.

  In Portugal, ERSAR has agreements in place with universities and knowledge centres and is a partner in national and international research and development projects, either promoting research and development or absorbing the knowledge it produces.

- *Studies on the sectors*: The regulator should conduct studies to improve knowledge of the sectors, addressing priority issues and delving further into technical aspects of interest to utilities providing water and waste services. It should also conduct studies on regulation of water and waste services and bring them to the knowledge of society and especially agents operating in the sectors.

  In Portugal, in recent years ERSAR has published a number of studies in the 'Estudos' series. They are available on its website (www.ersar.pt), in Portuguese:
  - *The regulation water supply and waste water treatment services – An international perspective, 2011.*
  - *History of public policies on water supply and treatment in Portugal, 2011.*

  It has also published articles on regulation, which are available on its website (www.ersar.pt):
  - *The strategic lines of the regulation model to be implemented by Instituto Regulador de Águas e Resíduos, 2003.*
  - *Contributions to the development of urban water and waste services in Portugal, 2003.*
  - *Recent developments and prospects for urban water and waste services in Portugal, 2005.*
  - *The quality of water for human consumption in Portugal: summary of the intervention of IRAR as the competent authority, 2005.*
  - *The need to invest in improving the quality of water for human consumption in Portugal. Reflections on the dispute with the European Commission, 2005.*
  - *The new Water Law and public water supply and waste water treatment services, 2006.*
  - *Criteria for access to the QREN (NSFR) 2007–2013 in the water and waste services sectors, 2006.*
  - *Status of the IRAR action plan for implementation of the PEAASAR II, 2007.*
  - *Status of the IRAR action plan for implementation of the PEAASAR II, 2008.*
  - *Brief history of the activity of the waters and waste regulation institute as the competent authority for quality of water for human consumption. Reflections and proposals for the future, 2008.*
  - *Challenges for water services in Portugal from a medium and long-term perspective, 2009.*
  - *The legal framework of water services in Portugal, 2010.*
  - *Model for calculating gains in productivity in the water and waste sectors in Portugal, 2010.*
  - *Study of economies of scale, scope and process in water and waste services, 2011.*
  - *Contributions to the reorganisation of the drinking water supply, waste water management and solid waste management services in Portugal, 2011.*
  - *Assessment of compliance of tariffs with tariff recommendations, 2012.*
  - *Tariffs for water for human consumption for non-domestic users, 2012.*
  - *The reform of the status and regulation of essential public services, 2013.*
  - *Water and waste services and their regulation, 2013.*

  In recent years it has also published a number of studies in 'Relatórios técnicos' series, which are available on its website (www.ersar.pt), in Portuguese:
  - *Quality of water from standpipes not connected to the public mains, 2005.*
  - *Delegation of powers of municipalities to parish councils and similar in the supply of water for human consumption, 2006.*

- *Public perception and willingness to pay for improvements in quality of water and waste services in mainland Portugal, 2007.*
- *Analysis of tariffs for solid waste management services in Portugal, 2007.*
- *Implementation of the polluter pays principle in the waste sector, 2013.*

- *Technical publications for the sectors*: The regulator must prepare technical publications for the practical problems specific to the sectors, such as technical manuals, to provide support to specialists at the utilities providing water and waste services, and technical courses for further training in specific areas of knowledge relevant to the sectors.

In Portugal, in recent years ERSAR has published a number of technical manuals in the 'Guias técnicos' series, often in collaboration with other entities. They are available on its website (www.ersar.pt), mostly in Portuguese:
- *Technical Guide 1: Performance indicators for water supply services, published by IRAR and LNEC, 2004.*
- *Technical Guide 2: Performance indicators for waste water services, published by IRAR and LNEC, 2004.*
- *Technical Guide 3: Control of losses from public water supply and distribution systems, published by IRAR, LNEC and INAG, 2005.*
- *Technical Guide 4: Modelling and analysis of water supply systems, published by IRAR and LNEC, 2006.*
- *Technical Guide 5 User manual for Epanet 2.0 – Simulations of water and quality parameters in water transport and distribution systems, published by IRAR and LNEC, 2004.*
- *Technical Guide 6: Control of quality of water for human consumption in public supply systems, published by IRAR, 2005.*
- *Technical Guide 7: Safety plans for water for human consumption, published by IRAR and Universidade do Minho, 2005.*
- *Technical Guide 8: Efficient water use in the urban sector, published by IRAR, LNEC and INAG, 2006.*
- *Technical Guide 9: Measuring flow in water supply and waste water treatment systems, published by IRAR and LNEC, 2006.*
- *Technical Guide 10: Operational control in public supply systems, published by IRAR, 2007.*
- *Technical Guide 11: Protection of surface and underground sources in water supply systems, published by IRAR and LNEC, 2009.*
- *Technical Guide 12: System for assessing quality of water and waste services provided to users – 1st generation of the quality of service indicator system, published by IRAR and LNEC, 2009.*
- *Technical Guide 13: Treatment of water for human consumption against quality of water at source, published by IRAR and LNEC, 2009.*
- *Technical Guide 14: Reuse of waste water, published by ERSAR and ISEL, 2010.*
- *Technical Guide 15: Solid waste management options, published by ERSAR, 2010.*
- *Technical Guide 16: Management of water supply infrastructures – A rehabilitation approach, published by ERSAR, LNEC and Instituto Superior Técnico, 2010.*
- *Technical Guide 17: Management of waste water and storm water infrastructures – A rehabilitation approach, published by ERSAR and LNEC, 2010.*
- *Technical Guide 18: Calculation of costs and earnings from water and waste services provided by utilities using a direct management model, published by ERSAR, 2012.*
- *Technical Guide 19: Manual on assessing the quality of water and waste services provided to users – 2nd generation assessment system, published by ERSAR and LNEC, 2012.*
- *Technical Guide 20: Utilities' relationship with users of water and waste services, published by ERSAR, 2012.*

In recent years ERSAR has also published a number of technical courses in the series 'Cursos técnicos', often in collaboration with other entities. They are available on its website (www.ersar.pt), in Portuguese:

  ○ *Technical Course 01: Urban hydrology – Basic concepts, published by ERSAR, Universidade de Coimbra and UNESCO, 2010.*

  ○ *Technical Course 02: Urban hydrology – Storm water drainage systems, published by ERSAR, Universidade de Coimbra and UNESCO, 2013.*

  ○ *Technical Course 03: Water Law, published by ERSAR and Faculdade de Direito – Universidade de Lisboa, 2013.*

  ○ *Technical Course 04: Waste Law, published by ERSAR and Faculdade de Direito – Universidade de Lisboa, 2014.*

- *Training courses*: The regulator should promote training courses for the sectors that address specific and relevant problems, to complement training at universities and other training centres, constituting a support instrument for specialists at utilities providing water and waste services. If there is a large number of regulated utilities and considerable geographical dispersal, the regulator should consider the use of e-learning for the purpose, at lower cost to the sectors.

  In Portugal, in recent years ERSAR has promoted a very large number of training courses for the sectors, sometimes in collaboration with other entities.

The regulatory activity described, following the regulatory procedures mentioned and the use of these regulatory instruments, allows the regulator to effectively and efficiently achieve the goals of technical supporting capacity building in the sectors, one of the components of the regulation model for the drinking water supply, waste water management and solid waste management services.

9.5 REGULATORY SYNERGIES

Support for capacity building in the sectors articulates closely with other components of the regulation model, as all these components naturally benefit from this support for capacity building. Conversely, the information from the components of legal contractual and economic regulation, regulation of quality of service and water quality and interaction with users can result in decisions on action to be taken in the field of capacity building and innovation in the sectors, out of a possible need for research and development, studies, manuals and technical courses or training in this area.

9.6 SUMMARY

This chapter gave a detailed description of one of the components of the proposed regulatory approach, the structural regulation component, which is called the contribution to capacity building of the sectors, including its objectives, activities, procedures, instruments and synergies.

The following chapter describes another of the components of the proposed regulatory approach, this time in the framework of regulation of the behaviour of the utilities, the legal and contractual regulation.

Chapter 10

Legal and contractual regulation

10.1 INTRODUCTORY NOTE

As part of the integrated regulatory approach (model RITA-ERSAR) proposed herein, this chapter describes in more detail the legal and contractual regulation of utilities, one of the components of the regulation model for the public services of drinking water supply, waste water management and solid waste management.

10.2 REGULATORY OBJECTIVES

This component of the regulatory model should aim to ensure that the entire life-cycle of the utilities, including the stages of design, tendering process, contracting, service management, contract amendment and termination, are carried out in strict compliance with legislation, which is monitored by the regulator, and also with any existing contract, as is the case in situations involving delegation and concessions.

This thereby helps to achieve the public policy goals, as specified in 3.3, of ensuring respect for the legislative framework for the public services involving the drinking water supply of water, waste water management and of solid waste management services to the population.

10.3 REGULATORY ACTIVITIES AND PROCEDURES

This regulatory component includes as its main activity the monitoring of the life-cycle of the utilities, but also includes the management of conflicts and the ongoing introduction of improvements to the sectors.

The regulator should monitor the utilities throughout their life-cycle, in the stages involving setting up new systems, the attribution of the management of services, day-to-day management of the service, alteration of contracts for management of services, approval of service regulations, celebration of contracts with users, authorisation to carry out secondary and supplementary activities, conflict resolution and termination of contracts for management of services. The analysis of the regulator is related not only to the legality of the elements analysed, but also to the technical and financial reasonableness of the specified assumptions. It is particularly intended to support the service holders in establishing tender and contractual conditions which enable it to take the best decision in the choice of the future delegate or concession holder, if this route is followed, optimising the conditions for operation of the service and safeguarding the rights of the utilities. In the event of analysing contract drafts, the focus of the analysis should be on whether it matches the tender documents and the winning bid, thus avoiding for example the redistribution of risks after the choice of the delegation or concession holder.

The regulator should periodically reassess and identify any need to promote the implementation of amendments to legislation, regulations or recommendations, based on the experience acquired with the ongoing monitoring of the life-cycle in terms of the legal and contractual compliance of the utilities. Such amendments shall be to address malfunctions and the improvement of the performance of the sectors and, subsequently, the service provided by the utilities.

Regarding the legal and contractual monitoring of utilities, the regulator should adopt a clear, specific and rational procedure as laid down in a regulatory procedural regulation, which ensures the efficient monitoring of each utility throughout their life-cycle.

There follows an example of a regulatory procedure that the regulator can follow in accompanying the life-cycle of the utilities:

- *The setting up of new systems*: Prior to the establishing of a water or waste system, the service holder should send the regulator the draft decision for establishing the respective system and any studies which substantiate this, particularly the economic and financial feasibility study. The decision may only be taken, otherwise it will be void, after the evaluation of the regulator, which should be mandatory but may or may not be binding, depending on the existing legislative framework.
- *Attribution of the management of services*: In the cases where the service holder intends to award the management of water or waste services to a legally distinct entity, the regulator should be asked in advance to make an evaluation on the formalisation of the act or contact which effects the transfer of responsibility for the management of the service in question, and such a request should be accompanied by elements which substantiate that decision from a legal, technical and economic point of view. The act or contract which affects the transfer of responsibility for management of the service may only be formalised after the issuing of the evaluation of the regulator, which should be mandatory but may or may not be binding, depending on the existing legislative framework. The regulator may also have the power to license the utility for the provision of water and waste services.

  A well-defined contract is essential and provides the means of controlling the delivery of the desired performance of a water and waste service utility, but it fails short of effective regulation. Effective regulation requires transparency, public participation and scrutiny of the contract.

  The regulation should be neutral regarding any preference for public or private, State or municipal management models, a decision which naturally is determined by the service holders, in accordance with legislation in force. The regulatory demands with regard to price and the quality of service provided to users should be the same, independently of the management model adopted. When the service holders decide to request the involvement of private utilities to provide a service for which they are incumbent by law, whether this be drinking water supply, waste water management or waste management, it is essential to safeguard the correct functioning of the competitive market which is thus created, as embodied in the public procurement procedures for the selection of this private utility, as well as its legitimate interests, particularly remuneration appropriate to the capital invested. The existence of a healthy market and a sustainable sectors also constitute guarantees for the provision of a quality service to users.

  In this context, and as far as the award and later management of delegations and concessions is concerned, the regulator should be especially attentive, prior to the opening of public procurement proceedings, to the most important aspects concerning the definition of the obligations of the future delegatees and concessionaires and the sharing of risks between the parties in particular. It should be a requirement that the attribution of these contracts be undertaken through public procurement procedures, with the assumption being that this will promote competition to enable the best solution to be chosen in the public interest. That is however competition by the market and not competition in the market.

- *Day-to-day service management*: To monitor the contracts, the regulator should carry out casuistic audits on the utility, whether the management is a direct, delegated or a concession management model. These actions seek to assess compliance not only with the legal norms related to the services, particularly the relationship with the users, but also with contracts, should these exist. The approach should not be strictly legal, but also economic, financial and technical, with the audit naturally being multidisciplinary. The process should include a first assessment of the information which the regulator has available concerning the utility to be audited, followed by a visit to the utility and the service holder to collect any additional information considered necessary. A preliminary report should be drawn up based on these elements, with a response period for one or for both parties, according to the specific case, which is converted after that into a final report. This final report should formulate recommendations to the parties to correct any non-conformities which have been detected and any possible penalties instigated, in accordance with the legislative framework.
- *Amendments of service management contacts*: Prior to any revision of contracts concerning water and waste services, the respective draft and annexes concerning the amendment should be sent to the regulator by the service holder for the issue of an evaluation seeking to assure the safeguarding of public interest and competition. The regulator, using the legal and contractual regulation module of its information system, should issue an evaluation with special concern for the maintenance of the essential assumptions of the contract which formed the basis for the choice of the delegatee or concessionaire and the distribution of risk between the parties, as established in that contract. The alteration should only be formalised, otherwise it will be void, after the evaluation of the regulator, which should be mandatory but may or may not be binding, depending on the existing legislative framework.
- *Approval of service regulations*: Prior to the approval of the service regulation, the utility should send the regulator the respective draft in order to issue its evaluation. The analysis should centre on its conformity with the legal norms which govern the relationship with the users, but also have as reference the need to ensure compliance with the provisions of any existing contract. The approval of service regulations may only be formalised, otherwise it will be void, after the evaluation of the regulator, which should be mandatory but may or may not be binding, depending on the existing legislative framework.
- *Celebration of contracts with the users*: Prior to celebrating any supply and collection contracts with the users of its services, the utility should send the regulator the respective drafts for the issue of an evaluation, using the legal and contractual regulation module of its information system. The contracts referred to in the previous paragraph may only be entered into after the issuing of an opinion regarding the respective draft contact. This analysis should centre on its conformity with the legal norms which govern the relationship with the users, but also have as reference the need to ensure compliance with the provisions of any existing contract. The celebration of contracts with the users may only be carried out, otherwise it will be void, after the evaluation of the regulator, which should be mandatory but may or may not be binding, depending on the existing legislative framework.
- *Authorisation for the carrying out of secondary and supplementary activities*: In the event of the carrying out of secondary and supplementary activities, the utility should request the evaluation of the regulator, which should be accompanied by proof of economic benefits that do not jeopardise the main activity and which also provide social, environmental or other benefits for the same,

particularly through an economic and financial projection of the activities in question. Authorisation for the carrying out of secondary and supplementary activities will only be granted, otherwise it will be void, after the evaluation of the regulator, which should be mandatory but may or may not be binding, depending on the existing legislative framework.

- *Resolution of conflicts*: The regulatory authority should intervene in the resolution of conflicts, particularly through mediation and conciliation, between any stakeholders subject to its intervention. These conflicts frequently result from different interpretations or non-compliances with the contracts by one or by both parties.

Figure 10.1 Legal and contractual regulation cycle.

- *Termination of service management contacts*: Prior to the termination of a contract concerning service management, the service holder should send the regulator the respective draft decision, duly substantiated, for the issue of an evaluation regarding the same, using the legal and contractual regulation module from the information system. The reasoning referred to in the previous paragraph should include, in the case of rescission, an indication of the contractual obligations not complied with by the utility and the respective consequences on the service provision to users. In the event of the redemption of the concession or the termination of the delegated management contract, this should involve the presentation of proof of the public interest that justifies this, and an explanation of the reasons particularly in terms of quality or efficiency of service provided, as well as the calculation of the due compensation owed to the concessionaire or any partner in the delegatee company. The decision to terminate contracts relating to the management of services may only be taken, otherwise it will be void, after the evaluation of the regulator, which should be mandatory but may or may not be binding, depending on the existing legislative framework.

The different stages for the regulation procedure for legal and contractual compliance are shown in Figure 10.1.

10.4 REGULATORY INSTRUMENTS

The regulator should promote, develop and use the most appropriate instruments for the regulation of the legal and contractual compliance of the utilities. Several examples of regulatory instruments will now be presented below:

- *Legislation*: To carry out this regulatory component, the regulator should follow the existing legislation for the sectors, as described above, which is essentially the legal framework of services, the legal framework on regulation, and regulations on tariffs, quality of service, water quality, technical requirements and regulatory procedures.
- *Utility contract*: In conducting this regulatory component, the regulator must naturally follow the contract that may exist between the service holder and the utility, which may for example be a delegated management contract between the service holder, as the delegator, and the utility, as the delegatee, and in a similar manner a concession contract between the service holder, as grantor, and the utility, as concessionaire, which define the relationship rules between the two parties.
- *Standard documents*: In addition, the regulator should make available standard-form documents concerning important aspects of the water and waste services management, which serve as a support for the utilities. Such forms are those referred to below:
  - *Standard-form for tendering procedures and delegated management contracts*: The availability of standard-forms for tendering procedures and delegated management contracts between the service holder, as delegator, and the utility, as delegatee, which are documents not necessarily binding that can be adopted and adapted to the specific needs of each case, facilitates the task of the service holder and permits, in particular, greater standardisation, compliance with legislation and good practices and safety in terms of contract specifications, the procedural schedule for the tendering procedure in the event of investment by a private partner, sharing of risks, criteria for assessing the bids and the actual text of the contract.
  - *Standard-form for tendering procedures and concession management contracts*: The availability of standard-forms for tendering procedures and concession management contracts between the service holder, as grantor, and the utility, as concessionaire, which are documents not necessarily binding that can be adopted and adapted to the specific needs of each case,

facilitates the task of the service holder and permits, in particular, greater standardisation, compliance with legislation and good practices and safety in terms of contract specifications, the procedural schedule for the tendering procedure, sharing of risks, criteria for assessing the bids and the actual text of the contract.

- ○ *Standard-form for service regulations*: The availability of standard-forms for service regulations, which are documents not necessarily binding that can be adopted and adapted to the specific needs of each case, facilitates the task of the utility and permits, in particular, greater standardisation, compliance with legislation and good practices and safety in terms of the definition of the rights and obligations of the utility and the users.
- ○ *Standard-form for user contracts*: The availability of standard-form contracts to enter into with users, which are documents not necessarily binding that can be adopted and adapted to the specific needs of each case, facilitates the task of the utility and permits, in particular, greater standardisation, compliance with legislation, service regulation and good practices and safety in terms of the definition of the rights and obligations of the utility and the users.

- *Inspections and audits*: The regulator should carry out inspections whenever it sees fit to do so as part of its powers of authority, to ensure ongoing assessment of utility compliance with legislation. It should also carry out audits whenever it sees fit to do so as part of its powers of authority, to ensure the reliability of the information supplied by the utility. Audits consist of a careful, systematic analysis of the activities carried out and information supplied by the utilities, in a scheduled or casuistic manner, with the goal of ascertaining if their activities are in conformity with legal and contractual provisions and if they are being implemented with effectiveness and efficiency in the attainment of their goals. The audits also determine the reliability of the information supplied. They may cover different areas, particularly legal and financial aspects and the quality of both of the service and the water.
- *Rules on regulatory procedures*: The regulator must make available rules on regulatory procedures and schedules that define in detail the procedures concerning relations between the regulator and the utilities under its regulation, as part of carrying out the duties and competences invested in that regulator by law, particularly in legal and contractual terms.
- *Conciliation processes*: The regulator should mediate and reconcile conflicts involving the utilities, by analysing these and fostering the use of conciliation and arbitration between the parties as a means of settling conflicts and taking any measures which it considers to be urgent and necessary.
- *Annual report:* The regulator should in an up-to-date annual report make a report on the legal and contractual compliance by the utilities, aimed at all stakeholders requiring reliable information, both to support and define policies and business strategies, as well as to assess the services actually provided to society.

In Portugal, ERSAR has published the annual report on water and waste services in Portugal (RASARP) since 2004. One of its sections deals with legal and contractual compliance by the utilities, and is available at its website (www.ersar.pt).

- *Information system*: In view of the large amount of information that this activity generates, regulatory effectiveness and efficiency can clearly be improved by using an information system as described above, particularly in relation to the regulation of legal and contractual compliance.

In Portugal, ERSAR is equipped with a sophisticated information system that is an indispensable instrument for its daily activity, which has various modules including one specifically for the regulation of legal and contractual compliance.

- *Penalties system:* There should be a fines system allowing the regulator to impose penalties on utilities for acts or omissions infringing legal or contractual provisions, particularly in relation to legal and contractual non-compliance.

The regulatory activity described, undertaken with the above-mentioned regulatory procedures and these regulatory instruments allows the regulator to effectively and efficiently achieve the goals of the regulation of the legal and contractual compliance of the utilities, one of the components of the regulation model for the public services of drinking water supply, waste water management and solid waste management.

10.5 REGULATORY SYNERGIES

The regulation of legal and contractual compliance articulates closely with other components of the regulation model. Firstly, the information gathered here can and should be cross-checked for validation with information arising from economic regulation, quality of service regulation, water quality regulation and user interaction. Moreover, the decisions contribute to better organisation of the sectors and information gathered in economic regulation and quality of service, quality of drinking water and interaction with users regulation can lead to intervention decisions in the regulation of legal and contractual compliance, such as the carrying out of audits to go more deeply into issues which have been identified.

10.6 SUMMARY

This chapter provided a detailed description of one of the components of the proposed regulatory approach, within the framework of the behavioural regulation of the utilities, designated as the regulation of legal and contractual compliance, including the respective objectives, activities, procedures, instruments and synergies.

The next chapter describes another of the components of the proposed regulatory approach, also within the framework of behavioural regulation of the utilities, the economic regulation.

Chapter 11

Economic regulation

11.1 INTRODUCTORY NOTE

As part of the integrated regulatory approach (model RITA-ERSAR) proposed herein, this chapter describes in more detail the economic regulation of the utilities, one of the components of the regulation model for drinking water supply, waste water management and solid waste management services.

11.2 REGULATORY GOALS

This regulatory model component has the aim of ensuring the application of tariffs suitable for the water and waste services, within a framework of economic and financial efficiency for the utilities, and also promotes the economic and financial sustainability of the utility and the suitability of the prices to the users' ability to pay.

Economic regulation is a way of regulating performances which is particularly linked to the regulation of quality of service, and constraining the permitted behaviour of the utilities regarding the tariffs they apply to users. It is also strongly influenced by some of the characteristics of the water and waste services, as these are high-value assets, which are long-lasting and involve high immobilisation, designed for peak situations, with long-term invested capital recovery and low elasticity between price and demand.

This also contributes towards the fulfilment of its public service obligations, as defined in 2.2, in terms of the adequacy of services pricing and fair prices for the services.

It also contributes towards achieving public policy goals, as defined in 3.3, thus ensuring the definition of a tariff policy and therefore providing the necessary financial resources, particularly through the construction and operation of water and waste services.

Given this, economic regulation should:

- Promote ongoing recovery of the service costs through tariff revenues (T1) to be paid by the users and ensure the sustainability of the utilities. Any tariff reduction through taxes (T2) from the national, regional or municipal budgets, or transfers (T3) of external funding support, should be a policy decision of the competent bodies to be taken only in exceptional situations. Subsidies result in general in unsustainable systems, and result in a lack of maintenance of the infrastructure with a downward spiral with deteriorating services. Additionally, they benefit the richer users more than the poor users. Subsidies to assist the poor should be directed to the poor' through the most focussed mechanisms possible.

- Promote more efficient tariffs, through incentives for operational efficiency, greater rationalisation in the choice of infrastructure construction, better use of idle capacity, effective management of infrastructure projects, where the utilities fulfil or even anticipate and go beyond the goals established for the sectors and the reduction of financial costs.
- Promote a suitable tariff structure, which should include the existence of an availability tariff, a fixed component, and a usage tariff, a variable component, with this latter being a progressive tariff consisting of four blocks, the first fulfilling the aim of social protection, the second having the main objective of recovery of costs, the third having the aim of being a small environmental penalty and the fourth having the aim of being a strong environmental penalty, to promote the efficient use of water.
- Safeguard macro and micro economic accessibility by final users, given that this is an essential service, so there should for example be a social tariff for low income families.

Economic regulation should be understood as an important mechanism for the behavioural regulation of utilities, in so far as activities carried out exclusively tend to generate inefficiency costs and lead to higher prices than those of a competitive market, therefore prejudicing the economic accessibility of users. Economic regulation should involve, on the one hand, a systematic assessment of the economic performance of the service utilities and an assessment of the reasonability of the tariffs applicable to final users, and, on the other hand, an assessment of investments by the utilities, insofar as these directly affect their economic future and financial sustainability. The interests of users are best served by a suitable selection of investment projects in the sectors and the appropriate financial coverage of these, essential aspects to ensure long-term continuity of service and maintenance of service levels in the short, medium and long term.

11.3 REGULATORY ACTIVITIES AND PROCEDURES

The main activities of this regulatory component include setting goals when carrying out the annual or multiannual regulation cycle and making on-going improvements in the sectors.

The regulator should ensure the existence of a tariff system, provided for both in legislation and regulation as well as in supplementary documentation, for example throughout tariff regulation, defining the criteria for tariff definition and the tariff structure, as well as the rules for the invoicing of services, which enable the utilities to define and carry out suitable management practices. This should ensure the principles of recovery of investment and operating costs, within a scenario involving efficiency, sustainable use of water resources, consumer protection interests and user's economic accessibility.

The regulator should annually or multi-yearly carry out an economic regulation cycle applicable to all the utilities providing public water and waste services using a set of procedures, presented in detail below.

Based on the experience acquired from the annual regulation cycle, the regulator should identify the need and if necessary work for the implementation of any methodological, procedural, legislative, regulatory or any other alterations defined in this regulatory component to improve the service provided by the utilities.

As mentioned the regulator should carry out an annual or multiannual economic regulation cycle applicable to all the utilities, and may use various economic regulation models to do this, involving greater or lesser regulatory intervention, as a function of the legal and contractual context, as presented below, which can be studied in detail on specialized books on economic regulation:

Indirect economic regulation

In this economic regulation model the tariffs practised by the utilities should be defined by them and naturally take into consideration the need to recover the costs incurred in providing the services within an efficiency scenario, including annual costs for the maintenance and substitution of infrastructure and equipment.

The regulator should intervene in the annual tariff revision cycle for services in a secondary level, solely with the aim of assessing the level of compliance with legislation, regulation and recommendations regarding tariffs, in accordance with the procedure described below.

The utilities should therefore focus on the calculation of costs and clarifying the calculation of tariffs, based on the principle of recovery of costs and productive efficiency, with it being necessary for these bodies to have the technical capacity to correctly calculate costs, check conformity with tariff regulation, assess the level of cost coverage, assess the level of economic accessibility of families and collect information on tariffs and disseminate this.

Economic regulation through contract

In this model of economic regulation the tariffs to be practised by the utilities result from provisions within their respective contracts, as a result of the bidding processes in which the winning proposals establish their tariffs in the first year and their formula for annual or multiannual updating is based on economic indicators.

The regulator intervenes in the annual or multiannual cycle for the tariff revision of services governed by delegation and the concession contract, essentially verifying compliance with the provisions for updating contracts, in accordance with the procedure described below.

Economic regulation through cost plus

In this model of economic regulation the tariffs practised by the utilities are regularly fixed, annually or multiannually, based on allowable established costs as stated in annual budgets, and as such it is designated as a cost plus model. Shareholder remuneration is contractually guaranteed, independently of the performance of the utility and is based on interest rates levied on the share capital and the legal reserve.

The regulator intervenes in the annual or multiannual cycle for analysing budgets and draft tariffs for services in accordance with the procedure described below, and it may define the tariffs or just provide an advice on them.

Its advantages are calculation simplicity, flexibility with regard to alterations of investment plans, which makes it suitable for contexts involving the extensive creation of infrastructure, and easy adaptation to alterations in the established presuppositions, which makes it suitable for contexts where sectors are not very mature.

Its disadvantages are the weak direct incentives for efficiency, in not allowing an effective transference of operational, financial and investment risks for the operator, its contribution towards a high level of tariff variation, as well as considerable pressure on tariffs following any alterations in its assumptions and the possibility of random investment decisions.

Economic regulation through revenue cap

In this model of economic regulation the tariffs practised by the utilities are regularly fixed, normally multiannual, based on the definition of the allowable incomes for the utilities, and as such are designated as a revenue cap model.

The revenue cap results from adding capital costs, with income and amortisation of investment in operating assets, operating costs incurred with the provision of the service in a scenario of productive efficiency and incentives to go beyond previously established goals, with any positive or negative adjustments resulting from variations in external factors. Income obtained by carrying out activities supplementary to the main ones, additional revenues resulting from activity and financial gains resulting from loans at a reduced or non-interest bearing rate are subtracted from these costs.

The tariff is naturally then that of the quotient of the revenue cap by the quantities envisaged for the service provided.

The regulator intervenes in the annual or multiannual cycle for defining the tariff for services in accordance with the procedure described below, and it may define the tariffs or only provide an advice on them.

Advantages are greater tariff stability, neutralisation of the impact on the utility of variations in demand and the utility taking on the operational, investment and funding risks. It also incorporates efficiency mechanisms at various levels of the tariff components.

Its disadvantages are greater calculation complexity, limited flexibility with regard to alterations of investment plans, which makes it suitable for contexts involving less intensive creation of infrastructure, and the difficult adaptation to alterations in the established assumptions, which makes it suitable for contexts involving mature sectors. It requires previously approved investment plans, take into account strategic national goals and also considering the existing installed capacity and the capacity necessary for the country.

Economic regulation through price cap

Regulation through price cap consists of establishing an average maximum ceiling for service prices during the regulatory period, and as such is designated as a price cap model. In this way, the utilities retain the profits corresponding to the reduction of costs which may incur during the regulatory period, in addition to envisaged productivity gains. At the end of each regulatory period the benefits of the reduction in costs are transferred to the users through a reduction in the prices for the following period.

The regulator intervenes in the annual or multiannual cycle to define the tariff for services in accordance with the procedure described below, and it may define the tariffs or only provide an advice on them.

Its advantages are that the regulators are given an incentive to increase efficiency, seeking to reduce costs and subsequently increase the respective revenue.

Its disadvantages are the greater calculation complexity, the limited flexibility with regard to alterations of investment plans, which makes it suitable for contexts involving less intense creation of infrastructure, and its difficult adaptation to alterations in the established assumptions, which makes it suitable for contexts involving mature sectors. It requires previously approved investment plans, take into account strategic national goals and also considering the existing installed capacity and the capacity necessary for the country.

Regarding economic regulation, the regulator should adopt a clear, specific and rational procedure for the model adopted, preferably as laid down in regulatory procedural rules, which ensures the efficient monitoring of each utility throughout the annual or multiannual life-cycle.

Given that tariffs may be annual or multi-annual, economic regulation intervention should be carried out for each utility at three distinct time periods for the event, which is the regulatory reference year or years. It should be carried out before the reference period (*ex-ante*) through specifying the rules and goals to be attained and approving the tariffs, during the reference period by applying the tariff and executing the economic and financial budget, and in the period following the reference period (*ex-post*) by collecting, validating and evaluating the economic and financial situation. This must be done at different levels, that of the utility, the region and the country, and, if required, adopting measures for later correction.

The procedures for the regulator regarding the various economic models considered are described below.

There follows a possible example of a regulatory procedure with regard to indirect economic regulation:

Before the annual or multiannual reference period (*ex-ante*):

- The regulator makes the tariff regulation available and publishes general recommendations for the purposes of tariff updating, including a forecast of macroeconomic indicators, in this way promoting standardisation of the tariff calculation by the utilities.
- After approval of the tariffs by the competent body, the utilities publicly disclose these and communicate them to the regulator, through the economic regulation module of the information system, along with the tariffs, the deliberations that approved them, as well as additional information to assess conformity with the tariff regulation.

During the annual or multiannual reference period:

- The utilities implement the approved budget and invoice users for their services based on the approved tariffs.
- The regulator uses audits to analyse the basis for the approved tariffs seeking to ascertain their level of compliance with tariff regulation and to receive recommendations, using the economic regulation module of the information system.
- In the case of non-compliance with the tariff regulation the regulator alerts the utility to the need to correct this or issues binding instructions in this regard.
- If justified, the regulator may open administrative infringement proceedings against the utilities and can apply penalties, under the terms set out in legislation.

At the end of each year (*ex-post*):

- The utilities submit, through the economic regulation module of the information system, their approved accounts and supplementary data which characterise economic and financial performance.
- The regulator validates this data in a first stage in the office and, where necessary, in a second stage through local audits, for the purposes of validating the economic and financial information and to obtain additional data.
- The regulator assesses the economic and financial performance of each of the utilities, and calculates the respective indicators.
- The regulator enables response processes with the utilities, through the economic regulation module of the information system, to enable the contradictory of its respective assessments of the economic and financial performance.
- The regulator benchmarks the utilities for each economic and financial indicator and for each cluster of utilities and assesses the evolution of these indicators over time at the level of each utility, the level of each cluster and at the national level.
- The cycle is concluded with the annual publication and public disclosure of the economic and financial assessment of the utilities by the regulator, highlighting the utilities with better performance, either through its annual report on water and waste services, or through other means of disclosure.

These different stages of the indirect economic regulation cycle are shown in Figure 11.1.

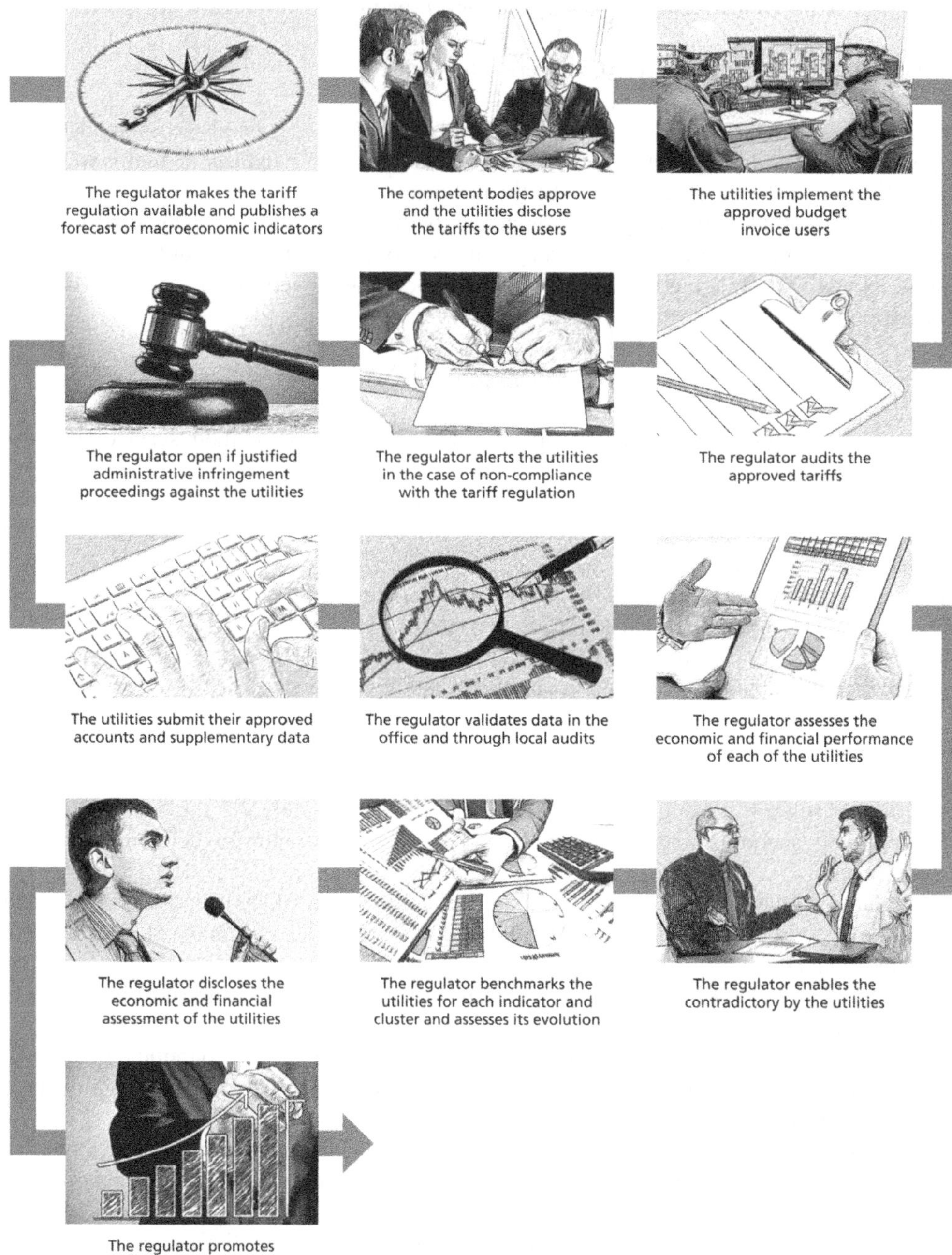

Figure 11.1 Indirect economic regulation cycle.

As far as economic regulation through contract is concerned, the following shows a possible example of regulatory procedure:

Before the annual or multiannual reference period (*ex-ante*):

- The regulator makes the tariff regulation available and publishes general recommendations for the purposes of tariff updating, including a forecast of macroeconomic indicators, in this way promoting standardisation of the tariff calculation by the utilities.
- The utilities send the regulator its tariff updating proposals using the economic regulation model of their information system, which should of course take into consideration legislative provisions, regarding tariff regulation, the contract and also the recommendations of the regulator.
- The regulator makes a preliminary evaluation on the proposals by the utilities for tariff updating, namely based on the contractual provisions and enables response processes, through the economic regulation module of the information system, and informs the service owner and the utility of this, so as to provide for the respective preliminary evaluations to be challenged.
- The regulator issues its definitive evaluation on the proposals by the utilities for tariff updating.
- After approval of the tariffs by the competent body, the utilities publicly disclose these and communicate them to the regulator, through the economic regulation module of the information system, along with the tariffs and the deliberations that approved them.

During the annual or multiannual reference period:

- The utilities execute the approved budget and invoice users for their services based on the approved tariffs.
- The regulator should verify through audits the actual application of the approved tariffs and, in the case of non-compliance, the regulator alerts the utility to the need to correct this or issues binding instructions in this regard.
- If justified, the regulator may open administrative infringement proceedings against the utilities and can apply penalties, under the terms set out in legislation.

At the end of each year (*ex-post*):

- The utilities submit, through the economic regulation model of their information system, their approved accounts and supplementary data which characterise economic and financial performance.
- The regulator validates this data at an initial stage in the office and, where necessary, in a second stage through local audits for the purposes of validating the economic and financial information and to obtain additional data.
- The regulator analyses the budgetary execution and assesses the economic and financial performance of each of the utilities, and calculates the respective indicators.
- The regulator enables response processes with the utilities, through the economic regulation module of the information system, to enable the contradictory of its respective assessments of the economic and financial performance.
- The regulator benchmarks the utilities for each economic and financial indicator and for each cluster of utilities and assesses the evolution of these indicators over time at the level of each utility, the level of each cluster and at the national level.
- The cycle is concluded with the annual publication and public disclosure of the economic and financial assessment of the utilities, highlighting the utilities with better performance, either through its annual report on water and waste services, or through other means of disclosure.

These different stages of the economic regulation cycle through contract are shown in Figure 11.2.

Figure 11.2 Economic regulation through contract.

There follows a possible example of a regulatory procedure with regard to direct economic regulation through cost plus:

Before the annual or multiannual reference period (*ex-ante*):

- The regulator makes the tariff regulation available and publishes general recommendations for the purposes of tariff updating, including a forecast of macroeconomic indicators, in this way promoting standardisation of the tariff calculation by the utilities.
- The utilities send the regulator its tariff updating proposals, using the economic regulation module of the information system, which should of course take into consideration legislative provisions, regarding tariff regulation, the contract and also the recommendations of the regulator, as well as including the necessary data for economic regulation through cost plus.
- The regulator analyses the proposal in detail in the light of the tariff regulation and its recommendations and makes a preliminary evaluation on the budgetary proposals and the draft tariffs of the utilities, based on the methodology of economic regulation through cost plus.
- The regulator enables response processes with the utilities, through the economic regulation module of the information system, and informs the service owner and the utility of this, making possible the contradictory.
- The regulator issues its definitive evaluation (binding or non-binding approval or opinion) on the proposals by the utilities for tariff updating.
- After approval of the tariffs by the competent body, the utilities publicly disclose these and communicate them to the regulator, through the economic regulation module of the information system, along with the tariffs, and, if applicable, the deliberations that approved them, as well as additional information to assess conformity with the tariff regulation.

During the annual or multiannual reference period:

- The utilities execute the approved budget and invoice users for their services based on the approved tariffs.
- The regulator should verify through audits the actual application of the approved tariffs and, in the case of non-compliance, the regulator alerts the utility to the need to correct this or issues binding instructions in this regard.
- If justified, the regulator may open administrative infringement proceedings against the utilities and can apply penalties, under the terms set out in legislation.

At the end of each year (*ex-post*):

- The utilities submit, through the economic regulation model of their information system, their approved accounts and supplementary data which characterise economic and financial performance.
- The regulator validates this data at an initial stage in the office and, where necessary, in a second stage through local audits for the purposes of validating the economic and financial information and to obtain additional data.
- The regulator analyses the budgetary execution and assesses the economic and financial performance of each of the utilities, and calculates the respective indicators.
- The regulator enables response processes with the utilities, through the economic regulation module of the information system, to enable the contradictory of its respective assessments of the economic and financial performance.
- The regulator benchmarks the utilities for each economic and financial indicator and for each cluster of utilities and assesses the evolution of these indicators over time at the level of each utility, the level of clusters and at the national level.

- The cycle is concluded with the annual publication and public disclosure of the economic and financial assessment of the utilities, highlighting the utilities with better performance, either through its annual report on water and waste services, or through other means of disclosure.

The graphic representation of these different stages of the economic regulation through cost plus is similar to that one presented to the economic regulation through contract.

There follows a possible example of a regulatory procedure with regard to direct economic regulation through revenue cap, similar to the cost plus procedure:

Before the annual or multiannual reference period (*ex-ante*):

- The regulator makes the tariff regulation available and publishes general recommendations for the purposes of tariff updating, including a forecast of macroeconomic indicators, in this way promoting standardisation of the tariff calculation by the utilities.
- The utilities send the regulator its tariff updating proposals, using the economic regulation module of the information system, which should of course take into consideration legislative provisions, regarding tariff regulation, the contract and also the recommendations of the regulator, as well as including the necessary data for economic regulation through revenue cap.
- The regulator analyses the proposal in detail in the light of the tariff regulation and its recommendations and makes a preliminary evaluation on the budgetary proposals and the draft tariffs of the utilities, based on the methodology of economic regulation through revenue cap.
- The regulator enables response processes with the utilities, through the economic regulation module of the information system, and informs the licensor or delegator of this, so as to provide for the respective preliminary evaluations to be challenge.
- The regulator issues its definitive evaluation (binding or non-binding approval or opinion) on the proposals by the utilities for tariff updating.
- After approval of the tariffs by the competent body, the utilities publicly disclose these and communicate them to the regulator, through the economic regulation module of the information system, along with copies of the tariffs, and, if applicable, the discussions that approved them, as well as additional information to assess conformity with the tariff regulation.

During the annual or multiannual reference period:

- The utilities implement the approved budget and invoice users for their services based on the approved tariffs.
- The regulator should verify through audits the actual application of the approved tariffs and, in the case of non-compliance, the regulator alerts the utility to the need to correct this or issues binding instructions in this regard.
- If justified, the regulator may open administrative infringement proceedings against the utilities and can apply penalties, under the terms set out in legislation.

At the end of each year (*ex-post*):

- The utilities submit, through the economic regulation model of their information system, their approved accounts and supplementary data which characterise economic and financial performance.
- The regulator validates this data at an initial stage in the office and, where necessary, in a second stage through local audits for the purposes of validating the economic and financial information and to obtain additional data.

- The regulator analyses the budgetary implementation and assesses the economic and financial performance of each of the utilities, and calculates the respective indicators.
- The regulator enables response processes with the utilities, through the economic regulation module of the information system, to enable the contradictory of its respective assessments of the economic and financial performance.
- The regulator benchmarks the utilities for each economic and financial indicator and for each cluster of utilities and assesses the evolution of these indicators over time at the level of each utility, the level of each cluster and at the national level.
- The cycle is concluded with the annual publication and public disclosure of the economic and financial assessment of the utilities, highlighting the utilities with better performance, either through its annual report on water and waste services, or through other means of disclosure.

The graphic representation of these different stages of the economic regulation through revenue cap is similar to that one presented to the economic regulation through contract.

There follows a possible example of a regulatory procedure with regard to direct economic regulation through price cap, similar to the cost plus procedure:

Before the annual or multiannual reference period (*ex-ante*):

- The regulator makes the tariff regulation available and publishes general recommendations for the purposes of tariff updating, including a forecast of macroeconomic indicators, in this way promoting standardisation of the tariff calculation by the utilities.
- The utilities send the regulator its tariff updating proposals, using the economic regulation module of the information system, which should of course take into consideration legislative provisions regarding tariff regulation, the contract and also the recommendations of the regulator, as well as including the necessary data for economic regulation through price cap.
- The regulator analyses the proposal in detail in the light of the tariff regulation and its recommendations and makes a preliminary evaluation on the budgetary proposals and the draft tariffs of the utilities, based on the methodology of economic regulation through price cap.
- The regulator enables response processes with the utilities, through the economic regulation module of the information system, and informs the service owner and the utility of this, so as to provide for the respective preliminary evaluations to be challenge.
- The regulator issues its definitive evaluation (binding or non-binding approval or opinion) on the proposals by the utilities for tariff updating.
- After approval of the tariffs by the competent body, the utilities publicly disclose these and communicate them to the regulator, through the economic regulation module of the information system, along with copies of the tariffs, the discussions that approved them, as well as additional information to assess conformity with the tariff regulation.

During the annual or multiannual reference period:

- The utilities implement the approved budget and invoice users for their services based on the approved tariffs.
- The regulator uses audits to analyse the basis for the approved tariffs seeking to ascertain their level of compliance with tariff regulation and to receive recommendations, using the economic regulation module of the information system.
- If justified, the regulator may open administrative infringement proceedings against the utilities and can apply penalties, under the terms set out in legislation.

At the end of each year (*ex-post*):

- The utilities submit, through the economic regulation model of their information system, their approved accounts and supplementary data which characterise economic and financial performance.
- The regulator validates this data at an initial stage in the office and, where necessary, in a second stage through local audits for the purposes of validating the economic and financial information and to obtain additional data.
- The regulator analyses the budgetary implementation and assesses the economic and financial performance of each of the utilities, and calculates the respective indicators.
- The regulator enables response processes with the utilities, through the economic regulation module of the information system, to enable the contradictory of its respective assessments of the economic and financial performance.
- The regulator benchmarks the utilities for each economic and financial indicator and for each cluster of utilities and assesses the evolution of these indicators over time at the level of each utility, the level of each cluster and at the national level.
- The cycle is concluded with the annual publication and public disclosure of the economic and financial assessment of the utilities, highlighting the utilities with better performance, either through its annual report on water and waste services, or through other means of disclosure.

The graphic representation of these different stages of the economic regulation through price cap is similar to that one presented to the economic regulation through contract.

Another important aspect of economic regulation is the approval of new investments for utilities, which must be based ideally on asset management planning to predict future investment needs with more sustainability.

When, due to exceptional and unforeseen reasons, it is confirmed that there is a need to carry out investments not envisaged in the contract or distinct from those foreseen, in cases involving delegated or concession management the utilities should obtain prior authorisation to undertake these.

The request should include the basis for the need and opportunity to carry out each one of the investments, an indication and justification of the amount of each of the investments and the envisaged schedule to carry these out, an incremental analysis of the impact of each of the investments on the average tariff, supported by suitable economic and financial projections, including the sources of finance, the debt service chart and the financial schedule of the total investment.

The decision to approve the new investment, particularly new infrastructure, should depend on an assessment of the reasonableness of the general technical solution proposed, as an added value for the system, its timeliness of execution and the reasonableness of the investment and operational costs and the respective conformity with the amount of the global authorised investment, according to the following criteria:

- A link should be made with the new infrastructure and the global plan. Whenever the design of the new infrastructure introduces significant alterations in the configuration of the system envisaged in the global plan, the technical or economic reasons which form the basis for the taking of the decision concerning the alterations should be assessed.
- The added value which the projected infrastructure confers on the global plan should be assessed, expressed for example in terms of the increase in population coverage, the improvement in the quality of service provided to users and/or the greater resilience of the system.
- The timeliness of carrying out the new infrastructure should be assessed from the perspective of logical time sequence planning for the construction of the overall system, that is, avoiding construction of infrastructure which has been unnecessarily brought forward.

- The reasonableness of the general technical solution for the proposed infrastructure should be assessed, within a framework of optimisation of costs. For example, when this concerns a treatment plant, compliance through this solution of the legal and technical requirements imposed by the environmental licensed authorities should be assessed.
- The reasonableness of the investment and operational costs budgeted for in the proposal should be assessed, particularly by comparing the respective unitary costs with those corresponding to other similar infrastructure. The reasonableness of the impact of the new investments on the tariff should also be assessed, supported by suitable economic and financial projections.

11.4 REGULATORY INSTRUMENTS

The regulator should promote, develop and use the most appropriate instruments for the economic regulation of the utilities. There follow some examples of regulatory instruments belonging to the regulator and external to the sectors:

- *Tariff system*: The regulator should essentially follow a tariff system applicable to the utilities by defining a set of principles, rules, instruments and indicators suitable to the existing reality. A possible tariff system and its various components will be presented below in greater detail, given the special importance for regulation, namely:
  - *General criteria for setting the tariff*: The regulator should have completely identified the tariff principles to be followed and the tasks associated with the setting of the tariffs, such as the assessment of costs, the assessment of necessary revenues, the definition of the tariff structure and the social concerns regarding the services.
  - *Criteria for establishing the tariff structure*: The regulator should avail itself of criteria to define the tariff structure which will serve as support for economic regulation.
  - *Criteria for invoicing services*: The regulator should avail itself of criteria to invoice the services which will serve as support for economic regulation.
  - *Economic and financial assessment indicators*: The use of these indicators should have the goal of determining a quantitative measurement of the efficiency and effectiveness of various economic and financial aspects of the service provided by the utilities, enabling data to evolve towards information and from this towards effective knowledge, which can particularly be used in regulatory terms.

    In Portugal, ERSAR has drawn up technical documents directly related to tariffs, available on its website (www.ersar.pt), particularly:

    - *Tariff Recommendation No. 1/2009: Formation of tariffs for end users of public services of drinking water supply, waste water management and solid waste management.*
    - *Technical Guide 18: Calculation of costs and earnings from water and waste services provided by utilities using a direct management model, 2012 edition.*

- *Tariff regulations*: The regulator should follow existing tariff regulations to carry out this regulatory component, which sets out the tariff system and specifies the rules to be followed by utilities in their service provision to users, which are naturally a function of the economic regulation model adopted.

  In Portugal, ERSAR approved the new tariff regulations for waste management services, applicable to all State-owned and municipality-owned utilities providing services, covering direct management, delegated management and concessioned management, available through its website (www.ersar.pt). ERSAR is currently drafting the new tariff regulations for water services.

- *Utility contract*: In conducting this regulatory component, the regulator must naturally follow the contract that may exist between the service holder and the utility, which may for example be a delegated management contract between the service holder, as the delegator, and the utility, as the delegatee, or, in a similar manner, a concession contract between the service holder, as grantor, and the utility, as concessionaire, which define the relationship rules between the two parties, particularly in tariff terms.

- *Inspections and financial audits*: The regulator should carry out inspections whenever it sees fit to do so as part of its powers of authority, to ensure ongoing assessment of utility compliance with legislation and regulations. It should also carry out audits whenever it sees fit to do so as part of its powers of authority, to ensure the reliability of the information supplied by the utility.

  In Portugal, ERSAR developed a standard inspection report, which covers the verification of all legal norms which may result in offences, the results of the meeting held with the utility, as well as a set of recommendations aimed at improving the performance of the utility.

- *Disclosure of reference cases*: The regulator should include the annual awarding of prizes within the assessment system for quality of service, in order to identify, reward and share reference cases concerning the public drinking water supply, waste water management and solid waste management services. Furthermore, this activity gives the opportunity to share specific cases where utilities constitute reference cases and to raise the awareness of utilities and the sectors in general with regard to issues of quality in system design, implementation, management and operation. There are of course many possible forms of awarding such a distinction, and this may for example involve the annual awarding of seals of approval and prizes for excellence. The seals may be awarded to a greater or lesser number of utilities providing services which, in the previous annual regulatory assessment period, complied with a set of previously established and highly demanding criteria. In addition, prizes for excellence may be awarded to a utility for each of the aforementioned areas which, in addition to meeting the criteria to be awarded a seal for quality of service, has shown outstanding performance or a remarkable improvement, particularly in economic and financial terms.

- *Annual report on water and waste services*: The regulator should in an up-to-date annual report make a report on the economic regulation of the utilities, aimed at all stakeholders requiring reliable information, both to support and define policies and business strategies, as well as to assess the services actually provided to society.

  In Portugal, ERSAR has published the annual report on water and waste services in Portugal (RASARP) since 2004. One of its sections deals with the economic and financial nature of the services provided by the utilities, and is available at its website (www.ersar.pt).

- *Information system*: The regulator should make use of an information system to increase its regulatory effectiveness and efficiency, with regard to the high volume of data which its activity creates, along with a set of applications which enable access by its technicians and by the utilities. This should include an economic regulation model.

  In Portugal, ERSAR provides a sophisticated information system, which is an indispensable instrument for its daily activity, which has various modules including one specifically for the assessment of quality of service.

- *Rules on regulatory procedures*: The regulator must have rules on regulatory procedures that define in detail the procedures concerning the relations of the regulator with the utilities under its regulation, as part of the duties and competences invested in it by law, particularly in economic terms.

- *Penalties system*: The regulator should utilise a suitable penalties system, which allows for the imposition of suitable penalties on utilities for acts or omissions infringing legal provisions, particularly in economic and financial terms.

The regulatory activity described, undertaken with the above-mentioned regulatory procedures and these regulatory instruments allows the regulator to effectively and efficiently achieve the goals of the economic regulation of the utilities, one of the components of the regulation model for the public services of drinking water supply, waste water management and solid waste management.

Some of these aspects will be discussed in more detail below.

11.5 TARIFF SYSTEM

11.5.1 Overview

As mentioned above, as far as the economic regulation to be carried out by the regulator, and taking into account the complexity of the issue, it is essential to make use of a suitable and well-defined tariff system, which is described below.

11.5.2 General criteria for setting the tariff

The drinking water supply, waste water management and solid waste management services are essential to the well-being of citizens, public health, economic activities and protection of the environment. Given this fact, it is recognised that the consumer has the right to access which tends towards the universal, as well as the continuity and quality of those services, within a framework of price efficiency and equity.

This means that they should pay the just price for these services, that is, which is sufficient to recover the costs of the utilities, which does not include inefficiencies and which is equitably spread among all users.

The services tariffs should be defined in such a way as to simultaneously safeguard the interests of users and the sustainability of the services. Revenue should be used only to finance their provision and not for other purposes, in order to ensure tariff moderation and avoid that the consumer of these services is not only paying for them but also for other activities by the utility in a rather non-transparent manner.

Each utility should start by setting the costs of these services, which will enable it to know the necessary revenues for their recovery. They should thus define a tariff structure which enables these revenues to be obtained, ensuring economic and financial sustainability but also introducing mechanisms for the social protection of users. Each one of these steps will now be described.

Definition of principles

The tariffs for water and waste services should respect the following principles:

- The principle of the recovery of costs, under the terms of which the tariffs for the water and waste services should enable an increasing recovery of the economic and financial costs resulting from their provision, in conditions which ensure the quality of the service provided and the sustainability of the utilities, operating within a scenario of efficiency in a way such as to not unduly penalise its users with costs resulting from any inefficient management of its systems.
- The principle of the sustainable utilisation of the water resources, under the terms of which the water services tariffs should contribute towards the sustainable management of the water resources

with increasing internalisation of the costs and benefits which are associated with their use, and penalising waste and excessively high consumption.

- The principle of prevention and recovery, under the terms of which the services tariffs for waste management should contribute towards the avoidance of and reduction in waste production, encouraging the adherence of final users to systems involving selective collection of materials and the recovery of waste.
- Principle of the defence of the interests of users, under the terms of which the tariffs should ensure the correct protection of the final user, avoiding possible abuses of a dominant position by the utility, on the one hand, with regard to continuity, quality and cost for the final user for the services provided and, on the other hand, with regard to the mechanisms for their supervision and control, which have been shown to be essential in situations involving monopoly.
- The principle of economic accessibility, under the terms of which the tariffs should correspond to the financial capacity of the final users, insofar as it is necessary to ensure almost universal access to the water and waste services.

In conformity with the principle of the recovery of costs, the tariffs for the water and waste services should, in particular, recover the following costs:

- Reintegration and amortisation, within a suitable period and in accordance with the applicable accounting practices, of the amount of the operating assets for the provision of services, resulting from investments carried out concerning the implementation, maintenance, modernisation, rehabilitation and the replacement of infrastructure, equipment or system-related means.
- The operational costs for the utility, particularly those incurred through the acquisition of materials and consumables, transactions with other utilities providing water and waste services, external supplies and services, including those resulting from allocating costs to services with activities and resources shared with other services carried out by the utility, or incurred through the remuneration of staff connected to its services.
- Financial costs attributable to the financing of services and, where applicable, suitable return on invested capital by the utility.
- The charges which are legally incumbent upon the provision of services, particularly those related to tax.

For the purposes of the recovery of costs, consideration should also be given to revenues other than tariffs, particularly contributions and non-repayable subsidies, in accordance with the period for the reintegration and amortisation of the assets resulting from subsidised investments, to operating subsidies which, due to exceptional reasons of a social nature, are associated with the provision of these services, and to other revenues associated with the provision of services or the making use of the resources associated with them.

The specific costs associated with the drainage of storm water and urban cleaning should ideally be excluded, respectively, from the series of costs to be recuperated through the tariff for waste water management and solid waste management, if necessary based on segregation or estimate, and should be recovered by revenues distinct from that of the service holders, such as through municipal taxes. The drainage of storm water and urban cleaning are clearly distinct services, of a communitarian nature and not individually indexable to each user.

The establishing of tariffs should avoid practices involving cross subsidisation between the different services and activities undertaken by the utilities, which can occur when the economic outturn generated by one or more activities is used in determining the price of the other. The tariffs for drinking water supply, waste water management and solid waste management should therefore recover the costs of each one of these three services and not create cross-subsidisation, contrary to practices which sometimes exist where supply tariffs are higher than necessary to compensate for the insufficiency of tariffs for waste water management and solid waste management. This results from the fact that supplying involves a product being delivered to the user who is therefore more willing to pay, whereas with waste water management and solid waste management other products are being taken from a user, who is therefore less willing to pay for this.

The tariffs for drinking water supply, waste water management and solid waste management may be differentiated according to whether final users are domestic or non-domestic. End users can be classified as the first kind who use urban buildings for residential purposes, and the rest being classified as non-domestic users. The State, local municipalities, autonomous funds and services which comprise the State-owned business sector and the local business sector should be considered as non-domestic final users. This differentiation corresponds to a form of protection for domestic users with regard to non-domestic ones, whose needs should be considered within the social and economic context. But the adequate balance of the differentiation for domestic users to non-domestic is always difficult to achieve.

Assessment of the costs of services

Providing these services requires high investment in constructing and renovating infrastructure and equipment, as well as significant operational costs.

An essential step in the definition of the tariff is for each utility to carry out a calculation of the costs of these services, which should include initial investments, investments for the substitution and expansion, maintenance, conservation and repair of all assets and equipment connected with the service, as well as the operation and efficient management of the resources utilised in carrying out the service.

Assessment of the revenues necessary for the services

The previously calculated charges have, of course, to be financed through revenues, which should ensure the coverage of costs for each utility with the aim of guaranteeing their sustainability and the maintenance quality of service provided to users, as well as the expansion and renovation of the systems and equity between generations, and therefore avoiding the burdening of future generations.

These revenues can be obtained through the choice of charging tariffs (T1), through the use of national, regional or municipal taxes (T2), or even transfers from abroad, for example community funds (T3). For the environmental preservation of water resources, the application of the principle of the user-payer through the first way, the charging of tariffs is normally provided for, which more effectively contributes towards raising the awareness of the consumer concerning a good use of resources. It is also a fairer solution from the intergenerational point of view, not transferring debt created through the operation of services in the present to the future. Subsidising the operation, for example through the municipal budget, should only be used in exceptional situations, when what is at stake is the economic accessibility of the users to the services, with suitable justification.

It is important to be aware that insufficiency of revenues regarding actual costs of services is clearly a false economy which can result in harm the users in the medium term. When revenues do not cover costs, and if there is no external form of subsidy, which burdens the citizens as taxpayers, there are only two solutions, either not providing the service, or lowering its quality, neither of which are acceptable in terms of public health and environmental protection.

Establishing the tariff structure

Having estimated the necessary revenues, it is necessary to establish the tariff structure. The existence of a fixed available component and another varying according to use has clearly been shown to be the fairest solution for users. Another solution may include the existence of just a variable component, of a necessarily higher value. This has the inconvenience of benefiting users with more than one dwelling and prejudicing users with a single dwelling, who in principle have fewer economic resources. A third solution may include the existence of just a fixed component, of a necessarily higher value, which clearly has the serious inconvenience of not reflecting the amount of volume utilised or produced by the consumer, encouraging waste and giving a completely wrong signal from the environmental point of view.

For these reasons, the two components should jointly be used in order to recover the costs from all the beneficiaries in a fairer manner, minimising social and environmental inconveniences. It should be underlined that the fixed component of the tariff should correspond to the charges which the utilities incur through making the service available, even if there is no use by the users.

The user tariff should be progressive, using blocks, for instance with a first block (5 m^3/month) having the main goal of social protection, with a second block (5 to 15 m^3/month) having as its main goal the recovery of costs, with a third block (15 to 25 m^3/month) also having as its main goal the recovery of costs but with some environmental penalty and with a fourth block (25 m^3/month) having the main goal of providing an environmental penalty.

Social concerns with the services

Finally, but importantly, the regulator should ensure that the whole population has access to these essential public services, particularly the lowest income population, through suitable mechanisms involving tariff moderation and guaranteeing economic accessibility. For this reason there should be a first domestic block, which is more accessible, and a social tariff, when the household has a gross income up to a certain amount. Of course, along with these measures, utilities should seek ongoing improvements in the efficiency of their services, and therefore reducing their costs.

Economic macroaccessibility of the users to these essential public services should be ensured, at the local or regional level, thus minimising the social impact of tariff updating.

The creation of a tariff standardisation instrument may also be necessary, for example a fund for tariff balancing, as a national or regional scheme for equalisation with the coverage and intensity considered necessary. In this way it will be possible to correct or at least minimise the inevitable difference between the unit costs of these services, which will tend to be low in areas of great population density, normally with higher average disposable incomes, and those which will tend to be high in areas with low population density, normally with lower average disposable incomes. Tariff balancing funds should however be used with extreme caution, since they may lead to the preservation of inefficiencies, as these may be subsidised with no incentive for their elimination.

Economic microaccessibility of the users to these essential public services should be ensured, at the household level, thus minimising the social impact of tariff updating in the most fragile and extreme social situations.

This measure is especially important within the framework of the United Nations having recognised that the drinking water supply and waste water management is a human right. This objective may be obtained through tariffs compatible with the economic capacity of the population, volumetric tariffs with progressive blocks, the removal of the automatic charge for contracting and water connection, social tariff and the removal of the guaranty for a deposit by the utility. Solutions involving positive discrimination with regard to the minority with difficulties in having access to water and waste services may even be considered with a view to providing universal access to the service.

The implementation of this measure can moderate the social impact of tariff correction, introducing greater fairness between users in protecting extreme social situations.

11.5.3 Criteria for establishing the tariff structure

As referred above, one important step is the establishing the tariff structure, which will be detailed here.

Having specified the revenues necessary for the utilities, it is then needed to adopt a tariff structure which not only enables these revenues to be generated but also create the social protection instruments and the economic incentives suitable to guide the behaviour of users.

The tariff structure is a set of tariffs applicable due to the service provision and their respective rules of application.

The tariff applicable to the user consists of a set of unitary values and other parameters and rules for the calculation which provides a way of determining the exact amount to be paid to the utility in return for the service provided.

Drinking water supply service

Tariffs for the water supply service should include a fixed component and variable component, so as to recover the costs from all users in an equitable manner.

Through applying supply tariffs, the utility should be obliged to carry out the following activities, and not individually invoice them:

- Supply of water;
- Operation, maintenance and renovation of service connections between the building system and the public system, up to a maximum extension for example of 20 m;
- Celebrating or altering a water supply contract;
- Making available and installing an individual meter;
- Making available and installing a totalising meter at the initiative of the utility;
- Regular scheduled readings and regular verification of the meter;
- Repair or substitution of meter, shut-off valve or cut-off valve, except for a reason attributable to the user.

Besides the supply tariffs mentioned it is considered possible, in contrast, that the utilities apply tariffs for services provided by the utility, related to its principal service, but which, due to their nature, particularly due to the fact that they are provided occasionally upon request by the users or a third party, which should be subject to specific invoicing. Auxiliary services may in particular include the following:

- Analysis of building and home supply installation projects;
- Carrying out of water connections, beyond the maximum extension for example of 20 m or where this operation is not the responsibility of the utilities;
- Carrying out building systems inspections at the request of users;
- Suspension of and reconnection of service due to non-compliance by the user;
- Suspension of and reconnection of service at the request of the user;
- Extra reading of water consumption following user request;
- Extra verification of meter at user request, except when its respective fault has been proved to be for a reason not attributable to the user;

- Temporary connection to the public system, particularly for building sites and works and temporary population concentrations, such as fairs, festivals and exhibitions;
- Information on the public supply system in the form of location plans;
- Supply of water in tankers, except when justified due to supply interruptions, particularly in situations where public health is at risk;
- Other services requested by the user, particularly repairs in a building or domestic supply system.

Water supplied for direct firefighting should not be subject to tariff, although it should be subject to measurement and estimation for the purpose of assessing the water balance of the systems.

Waste water management service

Tariffs for the waste water service should include a fixed component and variable component, so as to recover the costs from all users in an equitable manner.

Through applying waste water tariffs, the utility should be obliged to carry out the following activities, and not individually invoice them:

- Collection and handling of waste water;
- Operation, maintenance and renovation of service connections between the building system and the public system, up to a maximum extension for example of 20 m;
- Celebrating or altering a contract for waste water collection;
- Implementing and conserving connection boxes and their repair, except for a reason attributable to the user;
- Installation of an individual flow meter, when the utility has considered this to be technically and economically justifiable, and its substitution and maintenance, except for a reason attributable to the user;
- Regular scheduled readings and regular verification of the meter.

In addition to the waste water tariffs already mentioned, it must be admissible for the utilities to charge tariffs for auxiliary services, particularly for the following:

- Analysis of building and domestic sanitation systems;
- Carrying out of physical connections, beyond the maximum extension for example of 20 m or where this operation is not the responsibility of the utilities;
- Carrying out building and domestic sanitation system inspections or trials at the request of users;
- Clearing of building and domestic sanitation system;
- Extra verification of the flow meter at the request of the user, except when its respective fault has been proved to be for a reason not attributable to the user;
- Extra reading of waste water flow following user request;
- Transportation and final destination of sludge coming from septic tanks, collected through movable resources;
- Transportation and final destination of waste water, collected through movable resources;
- Information on the public sanitation system in the form of location plans;
- Other services requested by the user, particularly repairs in a building or domestic supply system.

As for the basis of the calculation of the waste water service tariffs, it should be considered that the volume of waste water collected corresponds to the product of the application of a value of 0.9 to the volume of water consumed.

Whenever the user does not have a supply service, the utility should estimate the respective consumption as a function of the average consumption of users with similar characteristics within the municipal area, as shown in the previous year. This method should also be applied when the user, making use of the supply service, has been shown to produce waste water from its own water.

Solid waste management service

Tariffs for the solid waste management service should include a fixed component and variable component, so as to recover the costs from all users in an equitable manner.

Through applying solid waste tariffs, the utility should be obliged to carry out the following activities, and not individually invoice them:

- Installation, maintenance and substitution of equipment for collection of solid waste and collection of specific integrated waste flows.
- Collection and handling of waste on a large scale, similar to that of solid waste, and small quantities of green waste coming from dwellings within the urban area.

In addition to the waste tariffs already mentioned, it must be admissible for the utilities to charge tariffs for auxiliary services, particularly for the clearing and washing of building pipes for waste discharge.

As for the base of the calculation of the solid waste management service, the quantity of waste subject to collection should be estimated using indicators from a specific base which show a significant statistical correlation with the actual production of solid waste by the final users, particularly water consumption, electricity consumption and the physical characteristics of urban buildings, such as their area or type, or be determined through weighing or volumetric systems whenever the utility understands this to be technically and economically viable.

As far as non-domestic users are concerned, it should also be possible to use as indicators various parameters associated with the type of activity carried out by the user, or carry out the direct determination of the quantity of solid waste which is collected based on specific weighing systems or volumetric systems, whenever this is shown to be technically and economically justifiable, through decision of the utility or at the request of the user.

The tariff structure should include special tariffs to cope with specific situations of a social, equitable or geographical character, with it being advisable to consider the following:

Social tariff

The tariffs for drinking water supply, waste water management and solid waste management should be reduced for final domestic users whose household has an overall gross income for the purposes of income tax on individuals which does not exceed a certain value, to be fixed by the service holder, which should not for example be greater than double the annual value of the guaranteed minimum monthly wage. This reduction, with regard to water services, should be carried out through the exemption of the fixed tariffs and the application to the total consumption of the user of variable tariffs for the first block, up to the monthly limit of 15 m^3 and, in the case of waste management services, through the exemption of the respective fixed tariff.

In this way it is ensured that the neediest users continue to have access to these essential public services, which constitute a human right, paying a symbolic value for essential consumption.

Public service tariff

The tariffs for drinking water supply, waste water management and solid waste management should equally be reduced for private social solidarity institutions, non-profit non-governmental organisations and other recognised charitable bodies whose social activity so justifies this. This reduction should not correspond to values lower than the tariffs applied by the utility to final domestic users.

Family tariff

The tariffs for drinking water supply, waste water management and solid waste management may also be reduced according to the composition of the family household for final domestic users. This reduction should be carried out through adjusting the consumption blocks with regard to the size of the household, under the terms specified by the service holder.

More than a social concern, this family tariff ensures equity amongst users, and avoids the fact that, for the same consumption per person, a small family pays through the tariff for the first block, while a large family pays through the tariff of the third or fourth block.

Seasonal tariff

The utilities should be able to differentiate tariffs as a function of the period of the year, when justifiable, so as to meet considerable fluctuations in demand which are seasonal for instance due to tourism or situations involving scarce water resources. This differentiation should be carried out through the alteration of the variable service tariffs, up to for example a limit of 30% of the values applied in the other periods, and the utility should ensure regular measurement of consumption.

11.5.4 Criteria for invoicing services

The invoice is an essential vehicle of communication in any commercial relationship, in particular within the framework of providing essential public services, which bring together the water and waste services, since it is through it that the utility makes its users aware of the service provided, the respective price and the necessary and useful information regarding the relationship established.

The tariffs should have a uniform structure for all the utilities, with the respective invoices for the water and waste services respecting the principle of transparency and be easy to understand for the final user, containing information on the utility and the user and specifying the services provided, the tariffs applied, the forms of payment and other relevant information.

Invoices should have a format and use of language which is simple and explicit and which facilitate their reading and the understanding of their content. In particular, the utility should not use acronyms in the invoice which make its understanding difficult and, if it does so, it should reserve a space to explain any concepts or acronyms present in the invoice.

When the drinking water supply, waste water management and solid waste management services are provided by different utilities, they should seek to establish agreements among themselves in order to present the final user with consolidated invoices and therefore generate economies of scope for the overall costs in the invoice processing of the various services.

Information common to the three services

The minimum information to appear on water and waste services invoices should include:

- *Data concerning the sending of the invoice*: name of the individual person for the designation of the legal person and the respective postal address or electronic address used for the purposes of sending the invoice;

- Identification of the final user, with the name of the individual person or legal person of the contract holder, taxpayer number, identification of the place where the services are provided, indication of the type of final user, that is to say whether domestic or non-domestic, and code number used by the utility for the quick identification of the final user in its customer management system.
- Identification and contacts of the body responsible for issuing the invoice, including its postal address and telephone and electronic contacts for the purposes of clarification of any issues relating to the invoice.
- Information concerning payment, with a total value to be paid or received, the deadline for payment, individual itemisation of the balance of the current account of the final user, in particular specifying previous unpaid bills, an indication of the number and amount in debt and identification of the available means of payment, including important information concerning their use.
- Invoice details, with number of the invoice or credit note (where applicable), date of issue, total amount for each service supplied without taxes, identification of other charges, taxes or services, the invoicing and charging of which has been carried out by the body issuing the invoice and the respective amounts and legal value added tax applicable to each service, the value of this tax and the total value of the invoice with tax.

Information regarding the public water and waste services should also include:

- Other contacts and operating hours of the user support services, in particular, places to attend the public, telephone call centre, fax line, telephone lines dedicated to specific issues, for example for communicating failures in supply and other ruptures in the public system, website and electronic addresses;
- References to authorise account direct debits;
- Space reserved for useful messages and the explanation of concepts and acronyms used in the invoice.

Information not related to the services provided should not be included in the invoice sent to the user, particularly that of a marketing nature, which can however be sent as an annex to the same.

Specific information concerning the drinking water supply service

The specific information concerning the use of the public drinking water supply service should, as a minimum, include the following:

- The starting and finishing dates of the period of service provision which is being invoiced, indicating the number of days that have passed;
- Nominal diameter of the installed water meter, and in the case of there being multiple meters installed this should indicate the virtual diameter;
- The two last actual readings carried out by the utility, their respective dates and average consumption ascertained in that period, expressed in m^3 per 30 days or litres per day;
- Indications of the period reserved for and the alternative means available for the user to communicate readings;
- Information regarding the quality of the water supplied, particularly through indicating the percentage of analyses regularly carried out and the percentage of analyses in compliance with their limiting values;
- Unitary value of the fixed supply tariff and the resulting value of its application to the period the invoice concerns;

- Indication of the method of calculating the volume of water consumed, particularly, whether as a measurement carried out by the utility, or by reading communicated by the user, or if this is an estimate by the utility;
- Volume of water consumed, divided into consumption blocks, where applicable;
- Unitary values of the variable applicable supply tariffs and value of the variable component resulting from its application to the consumption at each block, indicating any rectifications regarding volumes or values already invoiced
- Value of the water resource management tax attributable to the volume of water consumed, where applicable;
- Tariffs applied to any auxiliary supply services which have been provided.

Specific information concerning the waste water management service

The specific information concerning the use of the waste water management service should, as a minimum, include the following:

- Identification and contacts of the utility providing the waste water service, when distinct from the utility responsible for issuing the invoice;
- The starting and finishing dates of the period of service provision which is being invoiced, when distinct from the dates regarding the supply service;
- Unitary value of the fixed waste water tariff and the resulting value of its application to the period the invoice concerns;
- Indication of the method of calculating the effluent collected, particularly whether as a measurement carried out by the utility, or by indexing the volume of water consumed;
- Unitary values of the variable waste water tariff or the percentage applied to the amount invoiced for the supply of water, as applicable;
- Value of the variable waste water service component, discriminating any rectifications regarding volumes or values already invoiced;
- Value of the water resource management tax attributable to the volume of waste water collected, where applicable;
- Tariffs applied to any other auxiliary waste water services which have been provided.

Specific information concerning the solid waste management services

The specific information concerning the use of the solid waste management service should, as a minimum, include the following:

- Identification and contacts of the utility providing the solid waste management service, when distinct from the utility responsible for issuing the invoice;
- The starting and finishing dates of the period of service provision which is being invoiced, when distinct from the dates regarding the supply service;
- Unitary value of the fixed waste management tariff and the resulting value of its application to the period the invoice concerns;
- Indication of the method of applying the variable tariff for the waste management service, particularly whether through estimated measurement or indexing to a specific base indicator, which should be itemised;
- Value of the variable waste management service component, indicating any rectifications regarding volumes or values already invoiced;

- Value of the waste management tax attributable to the solid waste collected, where applicable.
- Tariffs applied to any auxiliary waste management services which have been provided.

Additional information

Information on the following issues, where applicable, should be provided as an annex to the invoice, on an annual basis, to the final users, regarding:

- A summary of the level of use of the services in the last 12 months, expressed in monetary and physical units, where applicable, showing average monthly values;
- The means of accessing detailed and updated information regarding the quality of service provided and the quality of water supplied;
- In the cases where the final user benefits from the application of a social tariff, information regarding the value which would have been invoiced under normal circumstances;
- Indication of the consequences of non-compliance with contractual obligations, particularly concerning not paying the invoice within the allotted time, including a description of the regime for the applicable interest on arrears;
- Information concerning environmental and civic awareness raising, covering good practices for the correct and efficient use of the services.

11.5.5 Economic and financial assessment indicators

The economic and financial indicators are a quantitative assessment measure of certain aspects of the utility. As a whole, the selected indicators should succinctly cover the most important economic and financial aspects in a way that is intended to be correct and balanced. Each indicator contributes to the quantification of performance of the utility within a given perspective, within a given area and for a given period of time, and facilitates the assessment of compliance with the goals and the analysis of its evolution over time. This thus simplifies an analysis which by its nature is complex.

An indicator should in itself contain important information, but it is inevitably a partial vision of the global reality of management of the utility and in general does not incorporate all of its complexity. As such, its use out of context may lead to incorrect interpretations. It is always necessary to analyse indicators as a whole, with knowledge of cause, and linked to their context.

The economic and financial assessment indicators are normally expressed by ratios between data from the utility. Each indicator is the result of a processing rule, which specifies all the data necessary for the calculation, the unit in which it should be expressed and the respective algebraic combination.

The regulator should define a set of economic and financial indicators in order to carry out an overall analysis of the economic and financial performance of the utilities which will suitably convey the situation. This assessment system enables data to evolve towards information and from this towards effective knowledge, which can particularly be used in regulatory terms.

In Portugal ERSAR has specified a set of economic and financial indicators for drinking water supply, waste water management and solid waste management, which are currently under revision.

11.5.6 Data necessary for assessment

Data naturally form the basic component for the construction of an economic and financial assessment system for drinking water supply, waste water management and solid waste management.

The utilities need to collect, compile and send a set of data regarding its activity to the regulator. Each item of data to be supplied by the utilities should comply with the definitions established by the regulator, refer to the period of time in which the assessment takes place, refer to the geographical area served with regard to the service being analysed and be as exact and reliable as possible.

Self-evaluation by the utilities of the quality of the database used to calculate the indicators is indispensable so that users of the information produced are aware of the confidence associated with them, thus avoiding wrong interpretations. The quality of the data to be supplied by the utilities to the regulator should therefore always be explained in terms of the accuracy of data and the reliability of the information source, as mentioned in chapter 12.

In Portugal ERSAR has specified a set of economic and financial data for drinking water supply, waste water management and solid waste management, which are currently under revision.

11.5.7 Reference values

The regulator should also define levels and bands for economic and financial indicators which are directly related to the wanted economic and financial sustainability of the utilities.

In Portugal ERSAR has specified a set of economic and financial levels and bands *for drinking water supply, waste water management and solid waste management, which are currently under revision.*

11.6 REGULATORY SYNERGIES

Economic regulation articulates closely with other components of the regulation model. Indeed the decisions taken should depend on information coming from legal and contractual regulation, quality of service regulation, water quality regulation and user interface regulation. It may also influence decisions contributing to better sectors organisation, to the possible need for strategic alterations, in clarifying the rules of the sectors, to the need to alter legislation, and the capacity building and innovation of the sectors, through the possible need to carry out studies or training in this area. This component also provides an important set of data for the regular production and disclosure of information concerning the sectors.

11.7 SUMMARY

This present chapter described in detail one of the components of the proposed regulatory approach, within the framework of the behavioural regulation of the utilities, designated as economic regulation, including its respective goals, activities and procedures, instruments and synergies.

The next chapter describes another of the components of the proposed regulatory approach, also within the framework of behavioural regulation of the utilities, the regulation of the quality of service provided to the users.

Chapter 12

Quality of service regulation

12.1 INTRODUCTORY NOTE

As part of the integrated approach (model RITA-ERSAR) being proposed, this chapter describes in more detail the quality of service regulation, one of the components of the regulation model for the public services of drinking water supply, waste water management and solid waste management.

12.2 REGULATORY GOALS

The quality of the service provided is an essential aspect, particularly in terms of user comfort and protection of public health and the environment. Reinforced by its characteristics as an irreplaceable and heterogeneous service located in space and time, the water quality supplied by the utilities should always be subject to suitable supervision, with this need reinforced by the fact that the service is provided as a natural monopoly.

This regulatory model component should therefore have the goal of ensuring the provision of suitable quality of service by the utilities to the users, under the terms of applicable legislation and the specifications of the regulator.

It thus helps to ensure compliance with public service obligations, in terms of universal access to services, adequacy of the quantity, quality and continuity of services and structural and operational efficiency of the utilities, as defined in 2.2.

It will also contribute to ensuring public policy goals for water and waste services in terms of access targets, quality of service goals and improvement in the operational efficiency of the utilities, as defined in 3.3.

It is however indispensable to start by defining what quality of service means, in general a vague and relatively indeterminate concept, which should be reflected in concrete measures of effectiveness and efficiency regarding specific aspects of the activity carried out and the performance of the utilities. Indicators which express the levels of the quality of service provided to users should be used, thus making the comparison between goals and results obtained direct and transparent, thereby simplifying a situation which otherwise would be too complex.

It should be the responsibility of the regulator to define the quality of service indicators, integrated within the assessment system, defined according to the country's context and its level of development, as well as defining reference values or brackets that reflect realistic objectives to be achieved by the utilities. The indicators are important regulation instruments, as they enable performance assessment based on clear definitions in a common language, as well as standardisation in the collection of information throughout the sectors.

The quality of service indicators may in general cover three major assessment areas, namely social sustainability, service management sustainability and environmental sustainability (Figure 12.1).

- *Indicators which reflect social sustainability*: this group of indicators is intended to assess the suitability of the user interface, that is, if the service provided to users in the year in which the assessment refers is suitable, with this being divided into aspects of service accessibility to users and the quality of service provided to users.
- *Indicators which reflect the sustainability of service management*: the aim with this set of indicators is to assess if the utility is carrying out basic measures to ensure that service provision is sustainable, subdividing this into aspects of the economic sustainability of the service, the infrastructural sustainability of the service and the productivity of its human resources.
- *Indicators which reflect environmental sustainability*: the aim with this set of indicators is to assess the level at which the environmental aspects associated with the activities of the utility are safeguarded, subdividing this into aspects of efficiency in the use of environmental resources and the prevention of pollution.

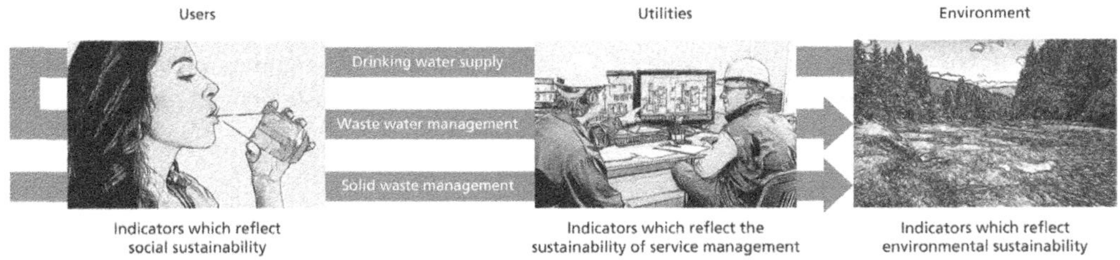

Figure 12.1 Groups of quality of service indicators.

In Portugal, ERSAR has defined, in partnership with the National Laboratory for Civil Engineering, one set of 16 indicators for each of the three services, which are described in detail in Technical Guide 19, entitled 'Guide for assessing the quality of water and waste services provided to users – 2nd generation assessment system', 2014 edition. This second generation corresponds to the evolution which has taken place in comparison with the first generation described in detail in Technical Guide 12, entitled 'System for assessing quality of water and waste services provided to users – 1st generation of the quality of service indicator system,' published in 2009 and which was in force for eight years. Both of these generations have been inspired by the IWA manuals on performance indicators, with regard to water supply and waste water services.

These indicators are annually calculated for close to 400 utilities of water and waste services, independently of their holding and management models, which may supply as a bulk, retail or integrated service. As an example, a public water system can be considered as usually having the bulk infrastructure components of abstraction, treatment, pumping and supply, with the retail infrastructure components being storage and distribution.

Given this, the following quality of service indicators are used for the water supply service:

Social sustainability

- *Physical accessibility of the service (%)*
- *Economic accessibility of the service (%)*
- *Occurrence of supply interruptions [No./(delivery point × year) for the utilities of bulk systems; No./(1000 connections × year) for the utilities of retail systems]*

- *Quality of supplied water (%)*
- *Reply to written suggestions and complaints (%)*

Service management sustainability

- *Coverage of total expenses (−)*
- *Effective connection to the service (%)*
- *Non-revenue water (%)*
- *Adequacy treatment capacity (%)*
- *Mains rehabilitation (%/year)*
- *Mains failures [No./(100 km × year)]*
- *Adequacy of human resources (No./(10^6 m^3 × year for utilities of bulk systems; No./1000 connections for utilities of retail systems)*

Environmental sustainability

- *Water losses [m^3/(km × day) for the utilities of bulk systems; l/(water connection × day) for the utilities of retail systems]*
- *Fulfilment of the water intake licensing (%)*
- *Energy efficiency of pumping installations [kWh/(m^3 × 100 m)]*
- *Disposal of sludge from the water treatment (%)*

The following quality of service indicators are used in the waste water management service:

Social sustainability

- *Physical accessibility of the service (%)*
- *Economic accessibility of the service (%)*
- *Flooding occurrence [No./(100 km of main pipe × year) for the utilities of bulk systems; No./(1000 connections × year) for the utilities of retail systems]*
- *Reply to written suggestions and complaints (%)*

Service management sustainability

- *Coverage of total expenses (−)*
- *Effective connection to the service (%)*
- *Adequacy of treatment capacity (%)*
- *Sewerage rehabilitation (%/year)*
- *Sewer collapses [No./(100 km × year)]*
- *Adequacy of human resources [No./(10^6 m^3 × year) for the utilities of bulk systems; No./(100 km × year) for the utilities of retail systems]*

Environmental sustainability

- *Energy efficiency of pumping installations [(kWh/(m^3 × 100 m)]*
- *Appropriate disposal of collected wastewater (%)*
- *Emergency control discharges (%)*
- *Wastewater analysis (%)*
- *Compliance with discharge parameters (%)*
- *Disposal of sludge from the wastewater treatment (%)*

The following quality of service indicators are used in the solid waste management service:

Social sustainability

- *Physical accessibility of the service (%)*
- *Accessibility of the separate collection service (%)*

- *Economic accessibility of the service (%)*
- *Container washing (−)*
- *Reply to written suggestions and complaints (%)*

Service management sustainability

- *Coverage of total expenses (−)*
- *Recycling of packaging waste (%)*
- *Organic recovery (%)*
- *Incineration (%)*
- *Landfill capacity usage (%)*
- *Renewal of the motor vehicle fleet (km/vehicle)*
- *Profitability of the motor vehicle fleet [kg/(m³ × year)]*
- *Adequacy of human resources (No./1000 t)*

Environmental sustainability

- *Use of energy resources (kWh/t for the utilities of bulk systems or tep/1000t for the utilities of retail systems)*
- *Quality of the leachate after treatment (%)*
- *Emission of greenhouse gases (kg CO_2/t)*

12.3 REGULATORY ACTIVITIES AND PROCEDURES

The regulator should ensure the existence of a suitable system to assess quality of service, as realised through both legislation and regulation, and also regarding any supplementary documentation, thus defining the assessment criteria and their reference values or levels, that is to say the quality of service goals to be attained.

Based on that system, it should annually carry out a quality of service regulation cycle applicable to all the utilities providing public water and waste services using a set of procedures.

In this cycle, which should take place in an on-going and planned manner, the regulator should follow a specific, clear and rational procedure, as laid down in a set of regulatory procedures. This regulatory activity should be carried out by each utility at two different periods regarding the regulatory year of reference, during the year of reference and after the year of reference (*ex-post*).

There follows an example of a regulatory procedure that the regulator can apply. It is divided into the following stages:

In the period corresponding to the reference year:

- The utility should carry out, during the year, ongoing collection of the internal and external data necessary to assess its quality of service, regarding the applicable indicators, as well as the definition of its profile and the system which it manages, as well as self-evaluation of the quality of the data in terms of its accuracy brackets and the reliability bracket of the information source, in accordance with the criteria defined.

In the period following the reference year (*ex-post*):

- The utility should send to the regulator, by the pre-defined date, based on the data relating to its activity of the preceding year and using the quality of service regulation module from the information system, the internal and external data necessary for its assessment, as well as its self-evaluation of the quality of that data in terms of the accuracy brackets and the reliability rating of the information source, as well as the files and the documentation needed for its respective validation.

- The regulator should, as a first step, carry out the validation of the results sent by the utility, and detect any data entry and processing errors and inconsistencies through cross analysis and validation of the data supplied and any clarifications in the event of doubts, using the quality of service regulation module from the information system.
- The regulator should, as a second step, carry out the validation of the data sent by the utility, and detect any data entry and processing errors, and verify its reliability, through a local audit of the utility, which should result in an audit report signed by the representatives of the regulator and endorsed by the utility.
- Once this step has finished, and until a pre-defined date, the regulator should process the data and proceed to interpret it through calculating the indicators, giving it for example meaningful symbols, for instance semaphoric codes, with regard to the defined reference values and intervals and the existing contextual factors, and it should also carry out an analysis and interpretation of its historical evolution, using the quality of service regulation module from the information system.
- The regulator should open a contradictory period, sending the utility and the service holder the respective indicators, contextual factors used and the preliminary interpretation of the results. It provides them with a pre-defined period to submit comments or suggest the correction of the assessment, suitably justified, and taking advantage of the opportunity to contest the assessment and allow for an additional analysis of the indicators and the contextual factors utilised, using the quality of service regulation module from the information system.
- Based on information which has been validated and responded to, the regulator should process the definitive data and interpret the results, with individual assessment of all the indicators for each utility with semaphoric codes, with global assessment for groups of utilities per each indicator, or benchmarking, also with semaphoric codes, and the assessment of the evolution of the indicators by each utility, region and country, using the quality of service regulation module from the information system.
- If justified, the regulator may open administrative infringement proceedings against the utility, under the terms set out in legislation, that is, if the information has not been reported or it has intentionally not represented the reality.
- The regulator should publicly disclose the results of this regulatory component through an annual report on water services, for professional use, and through interactive apps for non-professional use, on its website.

The different stages of the quality of service regulation cycle are shown in Figure 12.2.

Based on the experience acquired from the annual regulation cycle, the regulator should identify the need for improvements and if necessary work for the implementation of any methodological, procedural, legislative, regulatory or any other alterations to improve quality of service.

Benchmarking can be used to learn from the best utility, called process benchmarking. But it can be also used to describe systems of measuring relative performance of utilities, called metric benchmarking, which uses a relatively small number of measures compared with process benchmarking, which is much more detailed.

An effective metric benchmarking requires careful selection of the indicators to be used to direct utilities energies towards achieving the desired performance. There should be a few key indicators and a focus on obtaining reliable and meaningful data. Initially, effort should be focused on obtaining good data. Without meaningful data, it is not possible to measure progress of the utilities. Another issue is the clarity and soundness of the targets.

Figure 12.2 Quality of service regulation cycle.

Published benchmark tables of performance are a very useful incentive to management, but it is important that results are qualified to take into account the differences in context and operational conditions of each utility. Trends in performance of each utility, which show improvements over time, can be more meaningful than absolute performance figures.

12.4 REGULATORY INSTRUMENTS

The regulator should promote, develop and use the most appropriate instruments to regulate the quality of service of each utility. Various examples of regulatory instruments will be presented below:

- *Quality of service assessment system*: The regulator should essentially follow the assessment system for the service supplied by the utilities using a set of quality of service indicators regarding the drinking water supply, waste water management and solid waste management, suitable to the existing reality. The use of these indicators should have the goal of determining a quantitative measurement of the efficiency and effectiveness of various aspects of the service provided by the utilities. The quality of service assessment system enables data to evolve towards information and from this towards effective knowledge, which can particularly be used in regulatory terms. A quality of service assessment system and its various components will be presented below in greater detail, given its special importance for regulation.
- *Quality of service legislation*: The regulator should follow the existing legislation on quality of service to carry out this regulatory component which should set forth the rules to be followed by utilities in their service provision to users.

In Portugal, quality of service legislation is embodied in Decree-Laws No. 194/2009, of 20 August, and No. 95/2009, of 20 August, and Law No. 10/2014, of 6 March, which stipulate that the utilities should implement assessment mechanisms, the content of which includes, at least, a performance analysis system, and should use the model drawn up by the regulator.

Additionally, ERSAR has drawn up various technical guides related directly and indirectly to quality of service, also available on its website (www.ersar.pt):

- *Technical Guide 1: Performance indicators for water supply services, published by IRAR and LNEC, 2004.*
- *Technical Guide 2: Performance indicators for waste water services, published by IRAR and LNEC, 2004.*
- *Technical Guide 3: Control of losses in public water supply and distribution systems, published by IRAR and LNEC, 2005.*
- *Technical Guide 4: Modelling and analysis of water supply systems, published by IRAR and LNEC, 2006.*
- *Technical Guide 5: Epanet 2.0 user manual – Hydraulic simulation and quality parameters in water transportation and distribution systems, published by IRAR and LNEC, 2004.*
- *Technical Guide 8: Efficient use of water in the urban sector, published by IRAR and LNEC, 2006.*
- *Technical Guide 9: Flow measurement in water supply systems and waste water sewerage, published by IRAR and LNEC, 2006.*
- *Technical Guide 12: Assessment system for the quality of water and waste services provided to users – 1st Generation system and quality of service indicators, published by IRAR and LNEC, 2004.*
- *Technical Guide 14: Reuse of waste water, published by ERSAR and Instituto Superior de Engenharia de Lisboa, 2010.*
- *Technical Guide 15: Solid waste management options, published by ERSAR, 2010.*
- *Technical Guide 16: Management of water supply infrastructures – A rehabilitation-centred approach, published by ERSAR and LNEC, 2010.*
- *Technical Guide 17: Management of waste water and storm water infrastructure – A rehabilitation-centred approach, published by ERSAR and LNEC, 2010.*
- *Technical Guide 19: Assessment system for the quality of water and waste services provided to users – 2nd generation assessment system, published by ERSAR and LNEC, 2012.*

- *Quality of service regulations*: In addition to general legislation, the regulator should follow quality of service regulations, adding to legislation, defining the minimum levels of quality for services provided to users, and defining compensation due in cases of non-compliance with these minimum levels.

- *Disclosure of reference cases*: The regulator should include the annual awarding of prizes within the assessment system for quality of service, in order to identify, reward and share reference cases concerning the public drinking water supply, waste water management and solid waste management services. Furthermore, this activity gives the opportunity to share specific cases where utilities constitute reference cases and to raise the awareness of utilities and the sectors in general with regard to issues of quality in system design, implementation, management and operation. There are of course many possible forms of awarding such a distinction, and this may for example involve the annual awarding of seals of approval and prizes for excellence. The seals may be awarded to a greater or lesser number of utilities providing services which, in the previous annual regulatory assessment period, complied with a set of previously established and highly demanding criteria. In addition, prizes for excellence may be awarded to a utility for each of the aforementioned areas which, in addition to meeting the criteria to be awarded a seal for quality of service, has shown outstanding performance or a remarkable improvement.

  In Portugal, ERSAR promotes annually, together with a newspaper, some technical and scientific associations and a research center, the quality of service awards, recognizing the excellency of the nominated utilities.

- *Annual report*: The regulator should publish an up-to-date annual report on the situation regarding quality of service, aimed at all stakeholders requiring reliable information, both to support and define policies and business strategies, as well as to assess the services actually provided to society.

  In Portugal, ERSAR has published the 'Annual report on water and waste services in Portugal' since 2004. It includes a volume on the assessment of quality of service, and is available on its website (www. ersar.pt).

- *Inspections and audits*: The regulator should carry out inspections whenever it sees fit to do so as part of its powers of authority, to ensure ongoing assessment of utility compliance with legislation. It should also carry out audits whenever it sees fit to do so as part of its powers, to ensure the reliability of the information supplied by the utility.

  In Portugal, ERSAR developed a standard auditing report, which covers the assessing the quality of the water and waste services provided to users.

- *Information system*: The regulator should make use of an information system to increase its regulatory effectiveness and efficiency, with regard to the high volume of information which its activity creates, along with a set of applications which enable access by its technicians and by the utilities. This should include a quality of service module.

  In Portugal, ERSAR provides a sophisticated information system, which is an indispensable instrument for its daily activity, which has various modules including one specifically for the assessment of quality of service.

- *Penalties system*: The regulator should utilise a suitable penalties system, which allows for the imposition of suitable penalties on utilities for acts or omissions infringing legal provisions, particularly in terms of quality of service.
- *Rules on regulatory procedures*: The regulator must have rules on regulatory procedures that define in detail the procedures concerning relations with the utilities under its regulation as part of the duties and competences invested in it by law, particularly in terms of quality of service.

The regulatory activity described, undertaken with the above-mentioned regulatory procedures and these regulatory instruments, allows the regulator to effectively and efficiently achieve the goals for quality of service regulation, one of the components of the regulation model for water and waste services regulation.

12.5 QUALITY OF SERVICE ASSESSMENT SYSTEM
12.5.1 Overview

As mentioned above, as far as regulation of the quality of service of the utilities by the regulator, and taking into account the complexity of the issue, it is essential to make use of a suitable and well-defined assessment system for quality of service.

The most important components of this system are the quality of service indicators, which enable a quantitative measurement of efficiency and effectiveness to be determined for the various aspects of the service provided by the utilities and assess, in a quantified manner, compliance with the main goals laid down for this service. As a whole, the indicators should succinctly cover the most important aspects of quality of service in a way that is intended to be truthful and impartial. The quantification of each indicator, in contributing to the quantification of quality of service from a certain perspective, within a given area and for a given period of time, facilitates the assessment of compliance with the goals and the analysis of its evolution over time. The aim is thus to simplify an analysis which by its nature is complex.

Using quality of service indicators also provides for standardisation of information collection and performance assessment based on clear definitions using a common language.

The importance of the quality of service assessment system stems not only from being a powerful instrument to promote greater effectiveness and efficiency on the part of the utilities, but it also promotes an improvement in the quality of service which is supplied, as well as embodying a basic right of the users of these services, namely that of having access to reliable and easily interpreted information on the water and waste services being provided to them.

The quality of service assessment system has the following characteristics:

- The most important system component is naturally made up of the quality of service indicators, which are quantitative assessment measures of the efficiency and effectiveness of the service provided by the utility, described in detail below.
- The system is fed with data, which may be internal, relating to the utility and the system, and external, relating to the region where the utility is located.
- These data can evolve into information which describes the profile of the utility, important not only in the description it provides of the utility but also because it can provide support for the interpretation of the quality of service indicators. The profile description of the utility is described in detail below.
- These data can evolve into information which describes the profile of the system, important not only in the description it provides of the system but also because it can provide support for the interpretation of the quality of service indicators. The profile description of the system is described in detail below.
- These data can evolve into information which describes the contextual factors, important in providing support for the interpretation of the quality of service indicators. The contextual factors of the system are described in detail below.

- The system also requires reference values and brackets, specified by legislation, by the strategic plans or by the regulator, which should reflect the levels of service considered desirable for each quality of service indicator. In situations where the typology of the geographical area so justifies it, specific reference values and brackets should be defined according to area, for example for areas which are mainly urban, those moderately urban and those mainly rural. The reference values and brackets are described in detail below.

Assessment of the quality of the service is thus carried out based on the interpretation of the indicators, comparing the calculated values with the previously defined reference values and brackets, and attributing, for example, good, average or unsatisfactory performance, in accordance to the position within the reference values, in terms of being close or significantly distant from these. The interpretation should also receive additional support from information regarding the profile of the utility, the system profile and other contextual factors not included in the aforementioned profiles, which enable the interpretation to be customised for each situation.

Based on information which has been validated and responded to, the regulator should process the definitive data, and produce:

- An individual assessment of each utility for the various indicators with performance assessment, for example with meaningful symbols, according to the reference values and brackets.
- An assessment for each indicator at a national level and benchmarking comparing groups of similar utilities, also with performance assessment using meaningful symbols, for instance semaphoric codes, according to the reference values and brackets.
- Assessment of the historical evolution of each indicator at the national level, as regards groups of utilities and at the level of each utility.

The following figure shows the components of the quality of service assessment system and the flow of data which occurs, which is described in greater detail below (Figure 12.3).

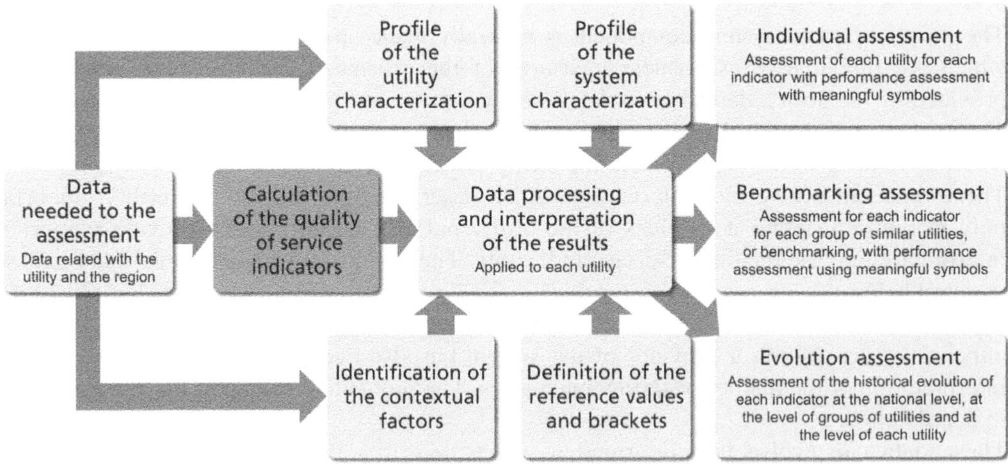

Figure 12.3 Quality of service assessment system.

Each of the components of the quality of service assessment system will now be presented in greater detail.

12.5.2 Quality of service indicators

The quality of service indicators adopted are quantitative assessment measures of efficiency and effectiveness for certain aspects of the service provided by the utility. Efficiency measures the extent to which available resources are used in an optimised manner for service creation. Effectiveness measures the extent to which realistically defined management objectives are met.

As a whole, the selected indicators should succinctly cover the most important aspects of quality of service in a way that is intended to be correct and balanced. Each indicator contributes to the quantification of performance within a given perspective, within a given area and for a given period of time, and facilitates the assessment of compliance with the goals and the analysis of its evolution over time. The aim is thus to simplify an analysis which by its nature is complex.

An indicator should in itself contain important information, but it is inevitably a partial vision of the global reality of management and in general, does not incorporate all of its complexity. As such, its use out of context may lead to incorrect interpretations. It is always necessary to analyse indicators as a whole, with knowledge of cause, and linked to their context.

Quality of service indicators are normally expressed by ratios between data from the utility. They may be undimensional, for example the data may be expressed in percentages, or dimensional, such as data expressed in euros per cubic metres. The relative denominator for the calculation represents one dimension of the system or utility being analysed, for example the number of service connections, the size of the pipes or collectors, the number of vehicles or annual costs. The use of denominators which may vary significantly from one year to another year due to factors external to the utility, for example annual water consumption, which can depend, among other factors, on the weather or other external factors, should be avoided as a denominator, unless the numerator varies in the same proportion.

Each indicator is the result of a processing rule, which specifies all the data necessary for the calculation, the unit in which it should be expressed and the respective algebraic combination. The data for the calculation of indicators may be directly generated and controlled by the utility, so-called internal data, or generated externally, so-called external data. In the system of indicators, the regulator should seek to minimise the calculation of quality of service indicators based on external data, since the utility has little room for manoeuvre regarding their quality control. In these cases it is preferable that the regulator takes on the direct collection of this data and its input into the system.

The system of indicators should be organised in accordance with the principles of the ISO 24500 standards, which specify that the performance objectives should be clearly identified, as well as the criteria to be adopted in assessing compliance with each objective and the performance indicators corresponding to each criterion.

Although the scope of these standards is limited to drinking water supply and waste water management services, the general principles enshrined may be equally generalised to solid waste management services.

In Portugal ERSAR, in its Technical Guide 19: Guide to assessing the quality of water and waste services provided to users – 2nd generation assessment system, published by ERSAR and LNEC, 2012, has specified the following quality of service indicators for water supply:

Social sustainability

- *Physical accessibility of the service (%): For utilities of bulk systems, the percentage of the total number of dwellings located in the area served by the utility for which there is bulk infrastructure that is connected or which can be connected to the retail system; for utilities of retail systems, the percentage of the total number of dwellings located in the area served by the utility for which there is infrastructure available for the water distribution service.*

- *Economic accessibility of the service (%): Weighting of the average water supply service cost on the average income available for each household in the area served by the system.*
- *Occurrence of supply interruptions [No./(delivery point × year) for utilities of bulk systems; No./(1000 connections × year) for utilities of retail systems]: For utilities of bulk systems, weighted average number of supply failures per delivery point, with the weighting factor the number of dwellings with an actual bulk service which depend on each delivery point; for utilities of retail systems, the number of supply failures per 1000 connections.*
- *Quality of supplied water (%) Percentage of controlled and good quality water, that is to say the product of the percentage compliance with sampling frequency and the compliance percentage with the values established in legislation.*
- *Reply to written suggestions and complaints (%): Percentage of written complaints and suggestions which received a written reply within 22 working days or less.*

Utility sustainability:

- *Coverage of total expenses (−): Ratio of total income and gains to total costs.*
- *Effective connection to the service (%) For utilities of bulk systems, the percentage of the total number of dwellings located in the area served by the utility in which the envisaged bulk infrastructure provides an effective service; for utilities of retail systems, the percentage of the total number of dwellings located in the area served by the utility for which there is infrastructure available for the water distribution service and this provides an effective service (with a water connection and contract even if temporarily suspended for a part of the year under analysis).*
- *Non-revenue water (%): Percentage of water which has entered into the system which has not been invoiced.*
- *Adequacy treatment capacity (%): Percentage of existing treatment capacity which was utilised in suitable conditions corresponding to its size.*
- *Mains rehabilitation (%/year): Annual average percentage of water supply and distribution main pipes more than ten years old which have been rehabilitated in the last five years.*
- *Mains failures [No./(100 km × year)]: Number of malfunctions in main pipes per unit length.*
- *Adequacy of human resources [No./(10^6 m³ × year) for utilities of bulk systems; No./1000 connections for utilities of retail systems]: For utilities of bulk systems, a number equivalent to the full time employees involved in the water supply service per unit volume of treated water exported; for the utilities of retail systems, a number equivalent to the full time employees involved in the water supply service per 1000 water connections.*

Environmental sustainability

- *Water losses [for the utilities of bulk systems, m³/(km × day); for the utilities of retail systems l/(water connection × day)]: For the utilities of bulk systems, actual water losses per unit of main pipe size; for the utilities of retail systems, volume of actual losses per water connection.*
- *Fulfilment of the water intake licensing (%): Percentage of the volume of abstracted water in licensed abstractions which meets the requirements of the abstraction use holders.*
- *Energy efficiency of pumping installations [kWh/(m³ × 100 m)]: Average standardised energy consumption of pumping stations.*
- *Disposal of sludge from the water treatment (%): Percentage of sludge from treatment system plants disposed of to a suitable destination.*

Some of those indicators may be considered as quality of service indicators of efficiency, measuring the extent to which available resources are used in an optimised manner for service provision: coverage of total expenses, non-revenue water, adequacy treatment capacity, mains rehabilitation, adequacy of human resources, water losses, and energy efficiency of pumping installations.

The others may be considered as quality of service indicators of effectiveness, measuring the extent to which management goals, realistically defined, are achieved: physical accessibility of the service, economic accessibility of the service, occurrence of supply interruptions, quality of supplied water, reply to written suggestions and complaints, effective connection to the service, mains failures, fulfilment of the water intake licensing, disposal of sludge from the water treatment.

In addition, ERSAR has defined the following quality of service indicators for waste water management:

Social sustainability

- *Physical accessibility of the service (%): For utilities of bulk systems, the percentage of the total number of dwellings located in the area served by the utility for which there is bulk infrastructure that is connected or which can be connected to the retail system; for utilities of retail systems, the percentage of the total number of dwellings located in the area served by the utility for which there is collection and drainage service infrastructure available for the water distribution service.*
- *Economic accessibility of the service (%): Weighting of the average waste water service cost on the average income available for each household in the area served by the system.*
- *Flooding occurrence [No./(100 km of main pipes × year) for the utilities of bulk systems; No./ (1000 connections × year) for the utilities of retail systems]: For the utilities of bulk systems, the number of occurrences of flooding on public highways and/or on properties originating from the public collection network, per 100 km of main pipes; for the utilities of retail systems, the number of occurrences of flooding on public highways and/or on properties originating from the public collection network, per 100 km of connections.*
- *Reply to written suggestions and complaints (%): Percentage of written complaints and suggestions which received a written reply within 22 working days or less.*

Sustainability of the entity

- *Coverage of total expenses (–): Ratio of total income and gains to total costs.*
- *Effective connection to the service (%) For utilities of bulk systems, the percentage of the total number of dwellings located in the area served by the utility in which the envisaged bulk infrastructure provides an effective service; for utilities of retail systems, the percentage of the total number of dwellings located in the area served by the utility for which there is infrastructure to provide access to the waste water service available and there is an effective service (with a water connection and contract).*
- *Adequacy of treatment capacity (%): Percentage of existing treatment capacity which was utilised in suitable conditions corresponding to its size.*
- *Sewerage rehabilitation (%/year): Annual average percentage of main pipes more than ten years old which have been rehabilitated in the last five years.*
- *Sewer collapses [No./(100 km × year)]: Number of structural collapses which have occurred per 100 km of main pipes.*
- *Adequacy of human resources [No./(10^6 m^3 × year) for utilities of bulk systems; No./(1000 km × year) for utilities of retail systems]: For utilities of bulk systems, this is equivalent to the number of full time employees involved in the waste water service per unit volume of waste water collected; for the*

utilities of retail systems, it is the number of full time employees involved in the waste water service per 100 km of main pipes.

Environmental sustainability

- *Energy efficiency of pumping installations [kWh/(m³ × 100 m)] Average standardised energy consumption of pumping stations.*
- *Appropriate disposal of collected wastewater (%) Percentage of the total number of dwellings located in the area served by the utility with a drainage service provided for by the public network and which are connected to a suitable treatment location.*
- *Emergency control discharges (%): Percentage of discharging with direct discharge to the monitored water source and which is functioning satisfactorily.*
- *Wastewater analysis (%): Percentage of the total number of analyses conducted compared to those required by the discharge licence or, in its absence, by applicable legislation.*
- *Compliance with discharge parameters (%): Percentage of the population equivalent which is serviced by treatment plants that ensure compliance with the discharge licence.*
- *Disposal of sludge from the wastewater treatment (%): Percentage of sludge from treatment system plants disposed of to a suitable destination.*

Some of those indicators may be considered as quality of service indicators of efficiency, measuring the extent to which available resources are used in an optimised manner for service provision: coverage of total expenses, adequacy of treatment capacity, sewerage rehabilitation, adequacy of human resources, and energy efficiency of pumping installations.

The others may be considered as quality of service indicators of effectiveness, measuring the extent to which management goals, realistically defined, are achieved: physical accessibility of the service, economic accessibility of the service, flooding occurrence, reply to written suggestions and complaints, effective connection to the service, sewer collapses, appropriate disposal of collected wastewater, emergency control discharges, wastewater analysis, compliance with discharge parameters, and disposal of sludge from the wastewater treatment.

In addition, ERSAR has defined the following quality of service indicators for solid waste management:

Social sustainability

- *Physical accessibility of the service (%): For utilities of bulk systems, the percentage of the solid waste collected in the area served by the utility which can enter the bulk processing infrastructure; for utilities of retail systems, the percentage of the total number of dwellings with a collection service at a distance of less than 100 m from the limit of the building (including door to door) in the area serviced by the utility.*
- *Accessibility of the separate collection service (%): Percentage of dwellings with a separate collection service through ecopoint bins (at a maximum distance of around 200 m) and/or door to door, made available by the utility in the area in which it operates.*
- *Economic accessibility of the service (%): Weighting of the average cost for the solid waste management service on the average income available for each household in the area served by the system.*
- *Container washing (–): Container washing frequency.*
- *Reply to written suggestions and complaints (%): Percentage of written complaints and suggestions which received a written reply within 22 working days or less.*

Sustainability of the entity

- *Coverage of total expenses (−): Ratio of total income and gains to total costs.*
- *Recycling of packaging waste (%): For utilities of bulk systems, the percentage of the packaging waste collected in the area served by the utility and reused for recycling; for utilities of retail systems, the percentage of the packaging waste selectively collected in the area serviced by the utility.*
- *Organic recovery (%) (just for the utilities of bulk systems): Percentage of waste subject to organic recovery within the area served by the utility.*
- *Incineration (%) (only for the utilities of bulk systems): Percentage of waste incinerated in the infrastructure of the utility itself within the area served.*
- *Landfill capacity usage (%) (only for the utilities of bulk systems): Percentage usage of the annual landfill capacity available in the infrastructure of the utility itself within the area served.*
- *Renewal of the motor vehicle fleet (km/vehicle): Average distance travelled by the vehicle allocated to the waste collection service.*
- *Profitability of the motor vehicle fleet [kg/(m³)] (only for utilities of retail systems): Quantity of waste collected divided by the annual installed capacity of collection vehicles.*
- *Adequacy of human resources (No./1000 t): For the utilities of bulk systems, total number of employees allocated to the waste management service per 1000 t of waste which entered the bulk processing infrastructure in the area served by the utility; for the utilities of retail systems: total equivalent number of full-time employees allocated to the solid waste management service per 1000 t of solid waste collected in the area serviced by the utility.*

Environmental sustainability

- *Use of energy resources (tep/t) (only for the utilities of retail systems): Total fuel consumption per 1000 tons of solid waste collected in the area served by the utility.*
- *Quality of the leachate after treatment (%) (only for the utilities of bulk systems): Percentage of the total number of analyses carried out on the treated leachate for which the results are in conformity with applicable legislation.*
- *Greenhouse gas emissions (kg CO_2/t): For the utilities of bulk systems, total quantity of CO_2 emissions from vehicles involved in selective collection of packaging per ton of collected waste in the area served by the utility; for the utilities of retail systems, total quantity of CO_2 emissions from collection vehicles per tonne of collected waste in the area served by the utility.*

Some of those indicators may be considered as quality of service indicators of efficiency, measuring the extent to which available resources are used in an optimised manner for service provision: coverage of total expenses, profitability of the motor vehicle fleet, adequacy of human resources, and use of energy resources.

The others may be considered as quality of service indicators of effectiveness, measuring the extent to which management goals, realistically defined, are achieved: physical accessibility of the service, accessibility of the separate collection service, economic accessibility of the service, container washing, reply to written suggestions and complaints, recycling of packaging waste, organic recovery, incineration, landfill capacity usage, renewal of the motor vehicle fleet, quality of the leachate after treatment, and emission of greenhouse gases.

The selection of proposed indicators should take into account requirements regarding each indicator, individually, and requirements relating to the set of indicators. Individually, each indicator requires:

- Rigorous definition, attributing concise meaning and unequivocal interpretation;
- Possibility to be calculated in total by the utilities without significant additional force;

- Possibility to be verified during audits;
- Simplicity and easy to interpret;
- Quantified, objective and impartial measurement on a specific aspect of quality of service, so as to avoid subjective or distorted judgements.

Collectively, the indicators should be defined so as to ensure the following requirements:

- Suitability regarding the representation of the main aspects of quality of service, thus enabling global representation;
- Absence of any overlapping meaning or objectives between indicators;
- Reference to the same period of time, with the assessment period adopted by the utility generally being that of one calendar year;
- Reference to the same geographical area, which should be clearly delimited and coincide with the area served by the utility regarding the service being analysed;
- Applicable to utilities with varied characteristics and level of development.

It should be noted that the quality of service indicators, in addition to their regulatory goals, are particularly important for the utilities themselves as a support instrument for system management, with a view to promoting ongoing improvement of the efficiency and effectiveness of the service. As such, it should be recommended to utilities that they use this instrument to assess compliance with their own management goals, not restricting themselves to quality of service indicators adopted for the purposes of regulation, but a more wide-ranging set considered important by each utility.

In Portugal, ERSAR has published the following technical guides, translated from two International Water Association publications, available on its website (www.ersar.pt), which present a large number of indicators which may be used by the utilities themselves:

- *Technical Guide 1: Performance indicators for water supply services, published by IRAR and LNEC, 2004, under authorization of IWA.*
- *Technical Guide 2: Performance indicators for waste water services, published by IRAR and LNEC, 2004, under authorization of IWA.*

12.5.3 Data necessary for assessment

Data naturally form the basic component for the construction of a quality of service assessment system for public drinking water supply, waste water management and solid waste management.

Indeed, to describe the profile of the utility and system profile, to calculate the quality of service indicators and define the contextual factors for the quality of service, the utility needs to collect, compile and send to the regulator a set of internal data regarding itself as a utility and the system which it operates. The regulator should supplement these data with any necessary external data.

Each item of data to be supplied by the utilities should be in accordance with the definitions published by the regulator, refer to the period of time in which the assessment takes place, refer to the geographical area served by the utility with regard to the service being analysed and be as exact and reliable as possible.

Self-evaluation by the utility of the quality of the database used to calculate the indicators is indispensable so that users of the information produced are aware of the confidence associated with them, thus avoiding wrong interpretations. The quality of the data to be supplied by the utilities to the regulator should therefore always be explained in terms of the accuracy of data and the reliability of the information source.

In accordance with metrological terminology, the accuracy of a measurement is its approximation between the result of the measurement and the conventionally true measured value. In this case, accuracy accounts for the error regarding the set of processes involving data acquisition and processing, including the error resulting from any extrapolation between certain measurements and the global value supplied. Given that, in general, it is not feasible to exactly know the error associated with each item of data, although it is much easier to know its order of magnitude, the accuracy of data should be communicated to the regulator in accordance with its classification bracket, as presented below:

Accuracy scale	Error associated with the data supplied
0–5%	Less than or equal to ± 5%
5–20%	Greater than ± 5%, but better than or equal to ± 20%
20–50%	Greater than ± 20%, but better than or equal to ± 50%
50–100%	Greater than ± 50%, but better than or equal to ± 100%
100–300%	Greater than ± 100%, but better than or equal to ± 300%
> 300%	Greater than ± 300%

This information may be used to estimate the accuracy level of the quality of service assessment indicators. Thus, assessment into the quality of the data should be complemented with an indication of the reliability of the information source, in accordance with the classification presented below:

Reliability bracket of the information source	Concept associated
***	Data based on exhaustive measurements, faithful recordings, procedures, research and investigations which are suitably documented and recognised as being the best calculation method.
**	Generically as before, but with some non-significant gaps in the data, such as part of the documentation being missing, calculations being old, or based on unconfirmed records, or also some extrapolated data having been included.
*	Data based on estimates and extrapolations from a limited sample.

The data to be annually supplied to the regulator by the utilities, which is necessary to describe the utility profiles and the system, and for calculating the quality of service indicators, naturally varies in accordance with the provision of drinking water supply, waste water management and solid waste management services.

In Portugal, ERSAR defined the following data for the assessment system for drinking water supply quality of service:

Identification of the utility

- *Identification of the utility (–)*
- *Model of governance (–)*
- *Bulk system(s) used (–) (only for the utilities of retail systems)*

- *Typology of the area served (–)*
- *Shareholder composition (–)*
- *Contract term (–)*

Dwellings

- *Dwellings with effective service (No.)*
- *Dwellings with non-effective available service (No.)*
- *Existing dwellings (No.)*

Complaints

- *Complaints and suggestions (No./year)*
- *Replies to complaints and suggestions (No./year)*

Failures and malfunctions

- *Failures in supply (No. failures × dwellings)/(delivery point × year) for the utilities of bulk systems; No./ year for the utilities of retail systems)*
- *Failures in main pipes (No./year)*

Water balance/water volumes

- *Water which entered into the system (m³/year)*
- *Authorised consumption (m³/year)*
- *Revenue water (m³/year)*
- *Non-revenue water (m³/year)*
- *Actual losses (m³/year)*
- *Abstracted water in licensed abstractions (m³/year)*
- *Abstracted water (m³/year)*
- *Treated water exported (m³/year)*

Control of water quality

- *Mandatory analyses carried out on water quality (No./year)*
- *Analyses carried out on parameters with a parametric value (No./year)*
- *Mandatory regulatory analyses on water quality (No./year)*
- *Analysis carried out in compliance with the parametric value (No./year)*

Energy

- *Energy consumption for pumping (kWh/year)*
- *Standardisation factor (m³/year × 100 m)*
- *Own production of energy (kWh/year)*
- *Energy consumption (kWh/year)*

Infrastructures and their use

- *Total size of main pipes (km)*
- *Average length of main pipes (km)*
- *Main pipes rehabilitated in the last five years (km)*
- *Connection pipes (No.) (only for the utilities of retail systems)*
- *Groundwater abstractions (No.)*
- *Surface water abstractions (No.)*
- *Pumping stations (No.)*
- *Water treatment plants (No.)*

- *Other treatment plants (No.)*
- *Storage tanks (No.)*
- *Reserve water capacity for supply and distribution (m³)*
- *Overuse of treatment plants (m³)*
- *Underuse of treatment plants (m³)*
- *Total capacity of treatment plants (m³)*
- *Management and infrastructure knowledge index (–)*

Treatment sludge

- *Sludge with suitable destination (t/year)*
- *Initial stored sludge (t/year)*
- *System produced sludge (t/year)*
- *Sludge from other systems (t/year)*
- *Final stored sludge (t/year)*

Economics

- *Income and total gains (€/year)*
- *Total costs (€/year)*
- *Average cost for water supply service (€/year)*
- *Average disposable family income (€/year)*
- *Tariff approved (€/m³) (only for the utilities of bulk systems)*

Certifications

- *Environmental management system certification (–)*
- *Quality management system certification (–)*
- *Occupational health and safety management system certification (–)*
- *Other certifications (–)*

Human resources

- *Water supply service staff (No.)*
- *Outsourced water supply service staff (No.)*

ERSAR has specified the following data for the assessment system for the waste water management quality of service:

Identification of the utility

- *Identification of the utility (–)*
- *Model of governance (–)*
- *Bulk system(s) used (–) (only for the utilities of retail systems)*
- *Typology of the area served (–)*
- *Shareholder composition (–)*
- *Contract term (–)*

Dwellings

- *Dwellings with effective service (No.)*
- *Dwellings with non-effective available service (No.)*
- *Dwellings serviced by individual waste water solutions (No.) (only for the utilities of retail systems)*
- *Dwellings with available untreated drainage (No.)*
- *Existing dwellings (No.)*

Equivalent population

- *Equivalent population with satisfactory treatment (valid discharge licence) (e.p.)*
- *Equivalent population with satisfactory treatment (expired discharge licence) (e.p.)*
- *Equivalent population serviced by treatment plants (e.p.)*

Complaints

- *Complaints and suggestions (No./year)*
- *Replies to complaints and suggestions (No./year)*

Failures and malfunctions

- *Flooding (No./year)*
- *Outflows functioning unsatisfactorily (No.)*
- *Structural collapses in main pipes (No./year)*

Control of water quality

- *Analyses requested (No./year)*
- *Analyses carried out (No./year)*

Waste water and energy

- *Invoiced waste water (m³/year)*
- *Collected waste water (m³/year)*
- *Volume of waste water treated and supplied to another body (m³/year)*
- *Volume of waste water treated and for own use (m³/year)*
- *Flow rate measurement index (−)*
- *Own production of energy (kWh/year)*
- *Energy consumption (kWh/year)*
- *Energy consumption for pumping (kWh/year)*
- *Standardisation factor (m³/year × 100 m)*

Infrastructures and their use

- *Total length of main pipes (km)*
- *Average length of main pipes (km)*
- *Main pipes rehabilitated in the last five years (km)*
- *Connection lines (No.) (only for the utilities of retail systems)*
- *Pumping stations (No.)*
- *Waste water treatment plants (No.)*
- *Collective septic tanks (No.)*
- *Treatment plants with valid discharge licence (No.)*
- *Overuse of treatment plants (m³)*
- *Underuse of treatment plants (m³)*
- *Total capacity of treatment plants (m³)*
- *Submarine outfalls (No.)*
- *Outflow pipes (No.)*
- *Unmonitored outflow pipes (No.)*
- *Management and infrastructure knowledge index (−)*

Treatment sludge

- *Sludge with suitable destination (t/year)*
- *Initial stored sludge (t/year)*

- *System produced sludge (t/year)*
- *Sludge from other systems (t/year)*
- *Final stored sludge (t/year)*

Economics

- *Income and total gains (€/year)*
- *Total costs (€/year)*
- *Average cost of waste water service (€/year)*
- *Average disposable family income (€/year)*
- *Tariff approved (€/m^3) (only for the utilities of bulk systems)*

Certifications

- *Environmental management system certification (−)*
- *Quality management system certification (-)*
- *Occupational health and safety management system certification (−)*
- *Other certifications (−)*

Human resources

- *Waste water staff (No.)*
- *Outsourced waste water staff (No.)*

Finally, ERSAR has specified the following data for the assessment system for the solid waste management quality of service:

Identification of the utility

- *Identification of the utility (−)*
- *Model of governance (−)*
- *System(s) user (−)*
- *Typology of the area served (−)*
- *Shareholder composition (−)*
- *Contract term (−)*

Dwellings

- *Dwellings with solid waste collection service (No.) (only for the utilities of retail systems)*
- *Dwellings with selective collection service (No.)*
- *Existing dwellings (No.)*

Complaints

- *Complaints and suggestions (No./year)*
- *Replies to complaints and suggestions (No./year)*

Quantity of waste

- *Solid waste collected (t/year)*
- *Waste which entered the bulk processing infrastructure (t/year) (only for the utilities of bulk systems)*
- *Solid waste which entered the bulk processing infrastructure (t/year) (only for the utilities of bulk systems)*
- *Packaging waste reused for recycling (t/year) (only for the utilities of bulk systems)*
- *Selectively collected packaging waste (t/year)*
- *Recycling volume (t/year)*
- *Solid waste subject to organic recovery (t/year) (only for the utilities of bulk systems)*
- *Solid waste directly disposed of in landfill (t/year) (only for the utilities of bulk systems)*

- *Waste subject to incineration (t/year) (only for the utilities of bulk systems)*
- *Solid waste disposed of in landfill (t/year) (only for the utilities of bulk systems)*
- *Non-solid waste disposed of in landfill (t/year) (only for the utilities of bulk systems)*
- *Waste to be deposited in landfill as envisaged in the licence (t/year) (only for the utilities of bulk systems)*
- *Solid waste collected (t/year) (only for the utilities of retail systems)*
- *Packaging waste reuse goal (t/year) (only for the utilities of bulk systems)*
- *Packaging waste collection goal (t/year) (only for the utilities of retail systems)*
- *Capacity to process biodegradable solid waste as defined in the strategic plan (t/year) (only for the utilities of bulk systems)*

Vehicles, equipment and their use

- *Distance travelled by the collection vehicles (km)*
- *Vehicles assigned to waste collection (No.)*
- *Installed capacity of waste collection vehicles (m^3/year) (only for the utilities of retail systems)*
- *CO_2 emissions for waste collection vehicles (kg CO_2)*
- *Number of washed containers (No./year)*
- *Number of containers (No./year)*

Quality of the leachate

- *Analyses requested for treated leachate (No./year) (only for the utilities of bulk systems)*
- *Analyses carried out on the treated leachate in accordance with legislation (No./year) (only for the utilities of bulk systems)*

Energy

- *Fuel consumption (tep/year) (only for the utilities of retail systems)*
- *Energy consumed by the external network (kWh/year) (only for the utilities of bulk systems)*
- *Sold energy obtained from energy recovery (kWh/year) (only for the utilities of bulk systems)*

Economics

- *Average cost of the solid waste management service (€/year)*
- *Average disposable family income (€/year)*
- *Income and total gains (€/year)*
- *Total costs (€/year)*
- *Tariff approved (€/m^3)* (only for the utilities of bulk systems):

Human resources

- *Waste management service staff (No.)*
- *Outsourced solid waste management staff (No.)*

Infrastructures

- *Eco-point bins (No.)*
- *Eco-centres (No.)*
- *Sorting stations (No.) (only for the utilities of bulk systems)*
- *Organic recovery units (No.) (only for the utilities of bulk systems)*
- *Incineration plants (No.) (only for the utilities of bulk systems)*
- *Landfills (No.) (only for the utilities of bulk systems)*
- *Transfer stations (No.)*
- *Installed incineration capacity (t/year) (only for the utilities of bulk systems)*
- *Installed container capacity (m^3) (only for the utilities of retail systems)*

Certifications

- *Environmental management system certification (–)*
- *Quality management system certification (–)*
- *Occupational health and safety management system certification (–)*
- *Other certifications (–)*

12.5.4 Profile of the utility

The profile of the utility is also an important component in the construction of an assessment system for the quality of service of public drinking water supply, waste water management and solid waste management.

The profile of the utility should be the set of features that summarily and uniquely characterise it. Besides the identification, the information should include aspects such as the model of governance, the system users, existing dwellings, the type of area served, the volume of activity, shareholder composition, the period of validity of the contract and existing certifications.

In Portugal, ERSAR defined the profile of the drinking water supply utility using the following aspects:

- *Identification of the utility (–): full official designation and address of the head office of the utility.*
- *Governance model (–): governance model adopted.*
- *Contract term (–): initial and final year of the period covered by the contract, where applicable.*
- *Shareholder composition (–): entities holding utility capital and their respective percentages, where applicable.*
- *Existing dwellings (No.): total number of dwellings in the area served by the utility for the water supply system.*
- *Activity volume (m³/year): total invoiced authorised consumption in retail systems and water invoiced in bulk systems (including exported water).*
- *Typology of the area served (–): classification of the area served by the utility regarding the typology of the urban area of the respective councils.*
- *Bulk system(s) used (–): bulk system(s) where the retail system is connected.*
- *Environmental management system certification (–): specification of certification regarding the water supply activity of the utility in accordance with ISO 14001 standard or similar.*
- *Certification of quality management systems (–): specification of certification regarding the water supply activity of the utility in accordance with ISO standard 9001 or similar.*
- *Occupational health and safety management system certification (–): specification of certification regarding the water supply activity of the utility in accordance with OHSA 18001 standard or similar.*
- *Other certifications (–): specification of other certifications regarding the utility's water supply activity.*

ERSAR has defined the profile of the waste water management utility by using the following aspects:

- *Identification of the utility (–): full official designation and address of the head office of the utility.*
- *Model of governance (–): model of governance adopted.*
- *Contract term (–): initial and final year of the period covered by the contract, where applicable.*
- *Shareholder composition (–): entities holding utility capital and their respective percentages, where applicable.*
- *Existing dwellings (No.): total number of dwellings in the area served by the utility for the waste water system.*
- *Dwellings serviced by individual waste water solutions (No.): number of dwellings located in the areas served by the utility where the collection and drainage infrastructure is connected and functioning, where there is waste water treatment.*

- *Activity volume (10^6 m³/year): invoiced waste water.*
- *Typology of the area served (–): classification of the area served by the utility regarding the typology of the urban area of the respective councils.*
- *Bulk system(s) used (–): bulk system(s) where the retail system is connected.*
- *Environmental management system certification (–): specification of certification regarding the waste water activity of the utility in accordance with ISO 14001 standard or similar.*
- *Certification of quality management systems (–): specification of certification regarding the water supply activity of the utility in accordance with ISO standard 9001 or similar.*
- *Occupational health and safety management system certification (–): specification of certification regarding the waste water activity of the utility in accordance with OHSA 18001 standard or similar.*
- *Other certifications (–): specification of other certifications regarding the utility's waste water activity.*

ERSAR has defined the profile of the solid waste management utility by using the following aspects:

- *Identification of the utility (–): full official designation and address of the head office of the utility.*
- *Model of governance (–): model of governance adopted.*
- *Contract term (–): initial and final year of the period covered by the contract, where applicable.*
- *Shareholder composition (–): entities holding utility capital and their respective percentages, where applicable.*
- *Existing dwellings (No.): total number of dwellings in the area served by the utility.*
- *Activity volume (t): for utilities of bulk systems, the quantity of waste which entered the processing infrastructure of the bulk utility; for utilities of retail systems, the total quantity of solid waste collected in the area serviced by the utility.*
- *Typology of the area served (–): classification of the area served by the utility regarding the typology of the urban area of the respective councils.*
- *Bulk system(s) used (–): bulk system(s) where the retail system is connected.*
- *Activity volume for recycling (t/year): for utilities of bulk systems, the quantity of solid waste sent for recycling; for utilities of retail systems, the quantity of solid waste selectively sent for recycling.*
- *Solid waste directly deposited in landfill (t/year): only for the utilities of bulk systems, the quantity of solid waste directly deposited in landfill.*
- *Environmental management system certification (–): specification of certification regarding the solid waste management of the utility in accordance with ISO 14001 standard or similar.*
- *Certification of quality management systems (–): specification of certification regarding the water supply activity of the utility in accordance with ISO standard 9001 or similar.*
- *Occupational health and safety management system certification (–): specification of certification regarding the solid waste management of the utility in accordance with OHSA 18001 standard or similar.*
- *Other certifications (–): specification of other certifications regarding the utility's solid waste management activity.*

12.5.5 System profile

The profile of the system is also an important component in the construction of an assessment system for the quality of service of the public drinking water supply, waste water management and solid waste management.

The profile of the system of the utility is understood to be the main characteristics which describe the set of infrastructures and support equipment for the service provided.

In Portugal, ERSAR defined the profile of the drinking water supply systems using the following aspects:

- *Surface water abstractions (No.): number of surface water abstractions under utility responsibility.*
- *Groundwater abstractions (No.): number of groundwater abstractions under utility responsibility.*
- *Water treatment plants (No.) number of water treatment plants under utility responsibility.*

- *Network length (km): total size of the water supply main pipes and distribution (not including branch connections).*
- *Density of connection branches (No. of connection /km network): number of existing connections per unit of length for the supply network.*
- *Pumping stations (No.): number of pumping stations under utility responsibility.*
- *Tanks (No.): Number of tanks of treated water under utility responsibility.*
- *Treated water reserve capacity (days): treated water supply autonomy for supply and distribution tanks.*
- *Own production of energy: percentage of energy consumed in which energy consumed is internally produced by the utility in the facilities allocated to the water supply service.*
- *Index of all infrastructural knowledge and management of patrimony (–): an index with values between 0 and 100 points, calculated according to information available on infrastructures, on interventions carried out and on the level of patrimony management.*

In addition, ERSAR has defined the profile of the waste water management systems using the following aspects:

- *Waste water treatment plants (No.): Number of waste water treatment plants under utility responsibility*
- *Total length of main pipes (km): total length of the main pipes managed by the utility.*
- *Submarine outfalls (No.): number of submarine outfalls under the utility responsibility.*
- *Pumping stations (No.): number of pumping stations under utility responsibility.*
- *Own production of energy (%): percentage of energy consumed in which the energy consumed is internally produced by the utility in the facilities allocated to the waste water service.*
- *Use of treated waste water (%): percentage volume of treated waste water that is used.*
- *Discharge licensing (%): percentage of waste water treatment plans with valid discharge licence.*
- *Index of all infrastructural knowledge and management of patrimony (–): an index with values between 0 and 100 points, calculated according to information available on infrastructures, on interventions carried out and on the level of patrimony management.*
- *Index of flow rate measurement (–): index with values between 0 and 100, calculated according to the existence of the flow rate measurement in treatment plants, weirs and bypass, pumping stations, in collection points, in the drainage network and alongside industrial users.*

Finally, ERSAR has defined the profile of the solid waste management systems by using the following aspects:

- *Ecopoints (No.): total number of existing ecopoints.*
- *Ecocentres (No.): total number of existing ecocentres.*
- *Vehicles assigned to waste collection (No.): number of vehicles assigned to solid waste collection.*
- *Sorting stations (No.): total number of existing sorting stations.*
- *Organic recovery units (No.) total number of existing organic recovery units.*
- *Incineration plants (No.): total number of existing incineration plants.*
- *Landfills (No.): total number of existing landfills.*
- *Transfer stations (No.): total number of existing transfer stations.*
- *Installed container capacity (m³): existing installed container capacity for the year under analysis.*

12.5.6 Contextual factors

The contextual factors are also important component in the construction of the quality of service assessment system for public drinking water supply, waste water management and solid waste management.

The contextual factors do not affect the assessment result, but may be taken into account by the regulator in the forming of its judgment, helping in the interpretation of some indicators and thus enabling a greater approximation to reality.

The utility and system profiles already take into account the main contextual factors that the regulator should generically take into consideration in this process. However, the assessment system for the quality of the service should foresee the possibility of including other contextual factors which were not considered from the outset. For this purpose, utilities should be able to identify, for any indicator, additional contextual factors which they consider determinant in the interpretation to be made by the regulator. The contextual factors to be specified should not be subject to any predefined format, but should always refer to auditable information.

12.5.7 Reference values

The regulator must specify quality levels and brackets for aspects which are directly related to the quality of service provided to users and experienced directly by them, as well as compensation due to these in the event of non-compliance.

*In Portugal, ERSAR defined the following levels and brackets for the quality of service indicators of **drinking water supply**:*

Physical accessibility of the service (%):

   For the utilities of bulk systems:

- *Good quality of service*   *100*
- *Average quality of service*   *[85; 100]*
- *Unsatisfactory quality of service*   *[0; 85]*

   For the utilities of retail systems:

- *Good quality of service*   *[95; 100]*
- *Average quality of service*   *[80; 95]*
- *Unsatisfactory quality of service*   *[0; 80]*

   (for mainly urban areas served)

- *Good quality of service*   *[90; 100]*
- *Average quality of service*   *[80; 90]*
- *Unsatisfactory quality of service*   *[0; 80]*

   (for moderately urban areas served)

- *Good quality of service*   *[80; 100]*
- *Average quality of service*   *[70; 80]*
- *Unsatisfactory quality of service*   *[0; 70]*

   (for mainly rural areas served)

Economic accessibility of the service (%)

   For the utilities of bulk systems:

- *Good quality of service*   *[0; 0.25]*
- *Average quality of service*   *[0,25; 0.50]*
- *Unsatisfactory quality of service*   *[0.50; +∞]*

For the utilities of retail systems:
- *Good quality of service*   *[0; 0.50]*
- *Average quality of service*   *[0.50; 1.00]*
- *Unsatisfactory quality of service*   *[1.00; +∞]*

Occurrence of supply interruptions [No./(delivery point) × year), for the utilities of bulk systems; No./ (1000 connections × year), for the utilities of retail systems]

For the utilities of bulk systems:
- *Good quality of service*   *0.00*
- *Average quality of service*   *[0.00; 0.20]*
- *Unsatisfactory quality of service*   *[0.20; + ∞]*

For the utilities of retail systems:
- *Good quality of service*   *[0.0; 1.0]*
- *Average quality of service*   *[1.0; 2.5]*
- *Unsatisfactory quality of service*   *[2.5; +∞]*

Quality of supplied water (%)

For the utilities of bulk and retail systems:
- *Good quality of service*   *[98.50; 100.00]*
- *Average quality of service*   *[94.50; 98.50]*
- *Unsatisfactory quality of service*   *[00.00; 94.50]*

Reply to written suggestions and complaints (%)

For the utilities of bulk systems:
- *Good quality of service*   *100*
- *Average quality of service*   *[95; 100]*
- *Unsatisfactory quality of service*   *[0; 95]*

For the utilities of retail systems:
- *Good quality of service*   *100*
- *Average quality of service*   *[85; 100]*
- *Unsatisfactory quality of service*   *[0; 85]*

Coverage of total expenses (-)

For the utilities of bulk and retail systems:
- *Good quality of service*   *[1.0; 1.1]*
- *Average quality of service*   *[0.9; 1.0] or [1.1; 1.2]*
- *Unsatisfactory quality of service*   *[0.0; 0.9] or [1.2; +∞]*

Effective connection to the service (%)

For the utilities of bulk systems:
- *Good quality of service*   *100*
- *Average quality of service*   *[98; 100]*
- *Unsatisfactory quality of service*   *[0; 98]*

For the utilities of retail systems:
- *Good quality of service*   *100,0*
- *Average quality of service*   *[90; 95]*
- *Unsatisfactory quality of service*   *[0; 90]*

Non-revenue water (%)

  For the utilities of bulk systems:
- *Good quality of service*   *[0.0; 5.0]*
- *Average quality of service*   *[5.0; 7.5]*
- *Unsatisfactory quality of service*   *[7.5; 100]*

  For the utilities of retail systems:
- *Good quality of service*   *[0.0; 20.0]*
- *Average quality of service*   *[20.0; 30.0]*
- *Unsatisfactory quality of service*   *[30.0; 100.0]*

Adequacy treatment capacity (%)

  For the utilities of bulk and retail systems:
- *Good quality of service*   *[90,100]*
- *Average quality of service*   *[70,90]*
- *Unsatisfactory quality of service*   *[0,70]*

Mains rehabilitation (%/year)

  For the utilities of bulk and retail systems:
- *Good quality of service*   *[1.0; 4.0]*
- *Average quality of service*   *[0.8; 1.0] or [4.0; 100]*
- *Unsatisfactory quality of service*   *[0.0; 0.8]*

Mains failures [No./(100 km × year)]

  For the utilities of bulk systems:
- *Good quality of service*   *[0; 15]*
- *Average quality of service*   *[15; 30]*
- *Unsatisfactory quality of service*   *[30; +∞]*

  For the utilities of retail systems:
- *Good quality of service*   *[0; 30]*
- *Average quality of service*   *[30; 60]*
- *Unsatisfactory quality of service*   *[60; +∞]*

Adequacy of human resources (No./(10⁶ m³ × year), only for the utilities of bulk systems; No./1000 branch connections, only for the utilities of retail systems)

  For the utilities of bulk systems:
- *Good quality of service*   *[1.0; 2.0]*
- *Average quality of service*   *[0.5; 1.0] or [2.0; 2.5]*
- *Unsatisfactory quality of service*   *[0.0; 0.5] or [2.5; +∞]*
  (for mainly urban areas served)
- *Good quality of service*   *[1.0; 2.5]*
- *Average quality of service*   *[0.5; 1.0] or [2.5; 3.3]*
- *Unsatisfactory quality of service*   *[0; 0.5] or [3.3; +∞]*
  (for moderately urban areas served)
- *Good quality of service*   *[1.0; 3.0]*
- *Average quality of service*   *[0.5; 1.0] or [3.0; 4.5]*
- *Unsatisfactory quality of service*   *[0; 0.5] or [4.5; +∞]*
  (for mainly rural areas served)

For the utilities of retail systems:
- *Good quality of service*   *[2.0; 3.0]*
- *Average quality of service*   *[1.5; 2.0] or [3.0; 3.5]*
- *Unsatisfactory quality of service*   *[0; 1.5] or [3.5; -]*

(for mainly urban areas served)
- *Good quality of service*   *[2.0; 3.5]*
- *Average quality of service*   *[1.5; 2.0] or [3.5; 4.3]*
- *Unsatisfactory quality of service*   *[0; 1.5] or [4.3; -]*

(for moderately urban areas served)
- *Good quality of service*   *[2.0; 4.0]*
- *Average quality of service*   *[1.5; 2.0] or [4.0; 6.0]*
- *Unsatisfactory quality of service*   *[0; 1.5] or [6.0; -]*

(for mainly rural areas served)

Water losses [l/(water connection × day)]

For the utilities of bulk systems:
- *Good quality of service*   *[0.0; 5.0]*
- *Average quality of service*   *[5.0; 7.5]*
- *Unsatisfactory quality of service*   *[7.5; +∞]*

For the utilities of retail systems:
- *Good quality of service*   *[0; 100]*
- *Average quality of service*   *[100; 150]*
- *Unsatisfactory quality of service*   *[150; +∞]*

Fulfilment of the water intake licensing (%)

For the utilities of bulk and retail systems:
- *Good quality of service*   *100*
- *Average quality of service*   *[90,100]*
- *Unsatisfactory quality of service*   *[0.90]*

Energy efficiency of pumping installations [kWh/(m³ × 100 m)]

For the utilities of bulk and retail systems:
- *Good quality of service*   *[0.27; 0.40] (average efficiencies between 68 and 100%)*
- *Average quality of service*   *[0.40; 0.54] (average efficiencies between 50 and 68%)*
- *Unsatisfactory quality of service*   *[0.54; +∞] (average efficiencies below 50%)*

Disposal of sludge from the water treatment (%)

For the utilities of bulk and retail systems:
- *Good quality of service*   *100*
- *Average quality of service*   *[95, 100]*
- *Unsatisfactory quality of service*   *[0, 95]*

*ERSAR has defined the following levels and brackets for the quality of service indicators of the **waste water management service**:*

Physical accessibility of the service (%):

For the utilities of bulk systems:
- *Good quality of service*   *100*
- *Average quality of service*   *[85, 100]*
- *Unsatisfactory quality of service*   *[0, 85]*

For the utilities of retail systems:
- *Good quality of service*   *[90; 100]*
- *Average quality of service*   *[80; 90]*
- *Unsatisfactory quality of service*   *[0; 80]*
(for mainly urban areas served)
- *Good quality of service*   *[85; 100]*
- *Average quality of service*   *[70; 85]*
- *Unsatisfactory quality of service*   *[0; 70]*
(for moderately urban areas served)
- *Good quality of service*   *[70; 100]*
- *Average quality of service*   *[60; 70]*
- *Unsatisfactory quality of service*   *[0; 60]*
(for mainly rural areas served)

Economic accessibility of the service (%):

For the utilities of bulk systems:
- *Good quality of service*   *[0; 0.25]*
- *Average quality of service*   *[0.25; 0.50]*
- *Unsatisfactory quality of service*   *[0.50; +∞]*

For the utilities of retail systems:
- *Good quality of service*   *[0; 0.50]*
- *Average quality of service*   *[0.50; 1.00]*
- *Unsatisfactory quality of service*   *[1.00; +∞]*

Flooding occurrence [No./(1000 connections × year), for the utilities of bulk systems; No./(1000 connections × year), for the utilities of retail systems]:

For the utilities of bulk systems:
- *Good quality of service*   *[0; 0.5]*
- *Average quality of service*   *[0.5; 2.0]*
- *Unsatisfactory quality of service*   *[2.0; +∞]*

For the utilities of retail systems:
- *Good quality of service*   *[0; 0.25]*
- *Average quality of service*   *[0.25; 1.0]*
- *Unsatisfactory quality of service*   *[1.0; +∞]*

Reply to written suggestions and complaints (%):

For the utilities of bulk systems:
- *Good quality of service*   *100*

- *Average quality of service*   *[95; 100]*
- *Unsatisfactory quality of service*   *[0; 95]*

For the utilities of retail systems:
- *Good quality of service*   *100*
- *Average quality of service*   *[85; 100]*
- *Unsatisfactory quality of service*   *[0; 85]*

Coverage of total expenses (-):

For the utilities of bulk and retail systems:
- *Good quality of service*   *[1.0; 1.1]*
- *Average quality of service*   *[0.9; 1.0] or [1.1; 1.2]*
- *Unsatisfactory quality of service*   *[0.0; 0.9] or [1.2; +∞]*

Effective connection to the service (%)

For the utilities of bulk systems:
- *Good quality of service*   *100*
- *Average quality of service*   *[90; 100]*
- *Unsatisfactory quality of service*   *[0; 90]*

For the utilities of retail systems:
- *Good quality of service*   *[99; 100]*
- *Average quality of service*   *[95.0; 99.0]*
- *Unsatisfactory quality of service*   *[0; 95]*

Adequacy of treatment capacity (%):

For the utilities of bulk and retail systems:
- *Good quality of service*   *[80; 100]*
- *Average quality of service*   *[60; 80]*
- *Unsatisfactory quality of service*   *[0; 60]*

Sewerage rehabilitation (%/year):

For the utilities of bulk and retail systems:
- *Good quality of service*   *[1.0; 4.0]*
- *Average quality of service*   *[0.8; 1.0] or [4.0; 100]*
- *Unsatisfactory quality of service*   *[0.0; 0.8]*

Sewer collapses [No./(100 km × year)]:

For the utilities of bulk systems:
- *Good quality of service*   *0.0*
- *Average quality of service*   *[0.0; 1.0]*
- *Unsatisfactory quality of service*   *[1; +∞]*

For the utilities of retail systems:
- *Good quality of service*   *0.0*
- *Average quality of service*   *[0.0; 2.0]*
- *Unsatisfactory quality of service*   *[2.0; +∞]*

Adequacy of human resources [No./(10^6 m^3 x year) for the utilities of bulk systems; No./(100 km × year) for the utilities of retail systems]:

For the utilities of bulk systems:
- *Good quality of service* [3.0; 4.0]
- *Average quality of service* [2.5; 3.0] or [4.0; 4.5]
- *Unsatisfactory quality of service* [0; 2.5] or [4.5; +∞]

(for mainly urban areas served)
- *Good quality of service* [3.0; 4.5]
- *Average quality of service* [2.5; 3.0] or [4.5; 5.3]
- *Unsatisfactory quality of service* [0; 2.5] or [5.3; +∞]

(for moderately urban areas served)
- *Good quality of service* [3.0; 5.0]
- *Average quality of service* [2.5; 3.0] or [5.0; 6.0]
- *Unsatisfactory quality of service* [0; 2.5] or [6.0; +∞]

(for mainly rural areas served)

For the utilities of retail systems:
- *Good quality of service* [5.0; 10.0]
- *Average quality of service* [2.5; 5.0] or [10.0; 12.5]
- *Unsatisfactory quality of service* [0; 2.5] or [12.5; +∞]

(for mainly urban areas served)
- *Good quality of service* [5.0; 11.0]
- *Average quality of service* [2.5; 5.0] or [11.0; 14.0]
- *Unsatisfactory quality of service* [0; 2.5] or [14.0; +∞]

(for moderately urban areas served)
- *Good quality of service* [5.0; 12.0]
- *Average quality of service* [2.5; 5.0] or [12.0; 15.5]
- *Unsatisfactory quality of service* [0; 2.5] or [15.5; +∞]

(for mainly rural areas served)

Energy efficiency of pumping installations [kWh/(m^3 × 100 m)]

For the utilities of bulk and retail systems:
- *Good quality of service* [0.27; 0.45] *average efficiencies between 60 and 100%)*
- *Average quality of service* [0.45; 0.68] *(average efficiencies between 40 and 60%)*
- *Unsatisfactory quality of service* [0.68; +∞] *(average efficiencies below 40%)*

Appropriate disposal of collected wastewater (%)

For the utilities of bulk and retail systems:
- *Good quality of service* 100
- *Average quality of service* [100; 95]
- *Unsatisfactory quality of service* [95; 0]

Emergency control discharges (%):

For the utilities of bulk and retail systems:
- *Good quality of service* [90; 100]

- *Average quality of service* *[80; 90]*
- *Unsatisfactory quality of service* *[0; 80]*

Wastewater analysis (%):

For the utilities of bulk and retail systems:
- *Good quality of service* *100*
- *Average quality of service* *[95; 100]*
- *Unsatisfactory quality of service* *[0; 95]*

Compliance with discharge parameters (%):

For the utilities of bulk and retail systems:
- *Good quality of service* *100*
- *Average quality of service* *[95; 100]*
- *Unsatisfactory quality of service* *[0; 95]*

Disposal of sludge from the wastewater treatment (%):

For the utilities of bulk and retail systems:
- *Good quality of service* *100*
- *Average quality of service* *[95; 100]*
- *Unsatisfactory quality of service* *[0; 95]*

ERSAR has defined the following levels and brackets for the quality of service indicators of **solid waste management service:**

Physical accessibility of the service (%):

For the utilities of bulk systems:
- *Good quality of service* *[95; 100]*
- *Average quality of service* *[80; 95]*
- *Unsatisfactory quality of service* *[0; 80]*

For the utilities of retail systems:
- *Good quality of service* *[95; 100]*
- *Average quality of service* *[80; 95]*
- *Unsatisfactory quality of service* *[0; 80]*
(for mainly urban areas served)
- *Good quality of service* *[90; 100]*
- *Average quality of service* *[80; 90]*
- *Unsatisfactory quality of service* *[0; 80]*
(for moderately urban areas served)
- *Good quality of service* *[80; 100]*
- *Average quality of service* *[70; 80]*
- *Unsatisfactory quality of service* *[0; 70]*
(for mainly rural areas served)

Accessibility of the separate collection service (%):

For the utilities of bulk and retail systems:
- *Good quality of service* *[90; 100]*
- *Average quality of service* *[70; 90]*
- *Unsatisfactory quality of service* *[0; 70]*
(for mainly urban areas served)

- *Good quality of service*   *[70; 100]*
- *Average quality of service*   *[50; 70]*
- *Unsatisfactory quality of service*   *[0; 50]*
(for moderately urban areas served)
- *Good quality of service*   *[50; 100]*
- *Average quality of service*   *[30; 50]*
- *Unsatisfactory quality of service*   *[0; 30]*
(for mainly rural areas served)

Economic accessibility of the service (%):

For the utilities of bulk systems:

- *Good quality of service*   *[0; 0.25]*
- *Average quality of service*   *[0.25; 0.50]*
- *Unsatisfactory quality of service*   *[0.50; +∞]*

For the utilities of retail systems:

- *Good quality of service*   *[0; 0.50]*
- *Average quality of service*   *[0.50; 1.00]*
- *Unsatisfactory quality of service*   *[1.00; +∞]*

Container washing (−):

For the utilities of bulk systems:

- *Good quality of service*   *[2; 6]*
- *Average quality of service*   *[1; 2] or [6; 12]*
- *Unsatisfactory quality of service*   *[0; 1] or [12; +∞]*

For the utilities of retail systems:

- *Good quality of service*   *[12; 24]*
- *Average quality of service*   *[6; 12] or [24; 30]*
- *Unsatisfactory quality of service*   *[0; 6] or [30; +∞]*

Reply to written suggestions and complaints (%):

For the utilities of bulk systems:

- *Good quality of service*   *100*
- *Average quality of service*   *[95; 100]*
- *Unsatisfactory quality of service*   *[0; 95]*

For the utilities of retail systems:

- *Good quality of service*   *100*
- *Average quality of service*   *[85; 100]*
- *Unsatisfactory quality of service*   *[0; 85]*

Coverage of total expenses (-):

For the utilities of bulk and retail systems:

- *Good quality of service*   *[1.0; 1.1]*
- *Average quality of service*   *[0.9; 1.0] or [1.1; 1.2]*
- *Unsatisfactory quality of service*   *[0.0; 0.9] or [1.2; +∞]*

Recycling of packaging waste (%):

For the utilities of bulk systems:

- *Good quality of service*   *[95; +∞]*

- *Average quality of service*   *[90; 95]*
- *Unsatisfactory quality of service*   *[0; 90]*

For the utilities of retail systems:

- *Good quality of service*   *[95; +∞]*
- *Average quality of service*   *[90; 95]*
- *Unsatisfactory quality of service*   *[0; 90]*

Organic recovery (%) (just for the utilities of bulk systems):

- *Good quality of service*   *[95; 100]*
- *Average quality of service*   *[90; 95]*
- *Unsatisfactory quality of service*   *[0; 90]*

Incineration (%) (only for the utilities of bulk systems):

- *Good quality of service*   *[75%; 100%] of installed capacity*
- *Average quality of service*   *[60%; 75%]of installed capacity*
- *Unsatisfactory quality of service*   *[0%; 60%] of installed capacity*

Landfill capacity usage (%) (only for the utilities of bulk systems):

- *Good quality of service*   *[0; 100]*
- *Average quality of service*   *[100; 110]*
- *Unsatisfactory quality of service*   *[110; +∞]*

Renewal of the motor vehicle fleet (km/vehicle):

For the utilities of bulk and retail systems:

- *Good quality of service*   *[0; 250 000]*
- *Average quality of service*   *[250 000; 350 000]*
- *Unsatisfactory quality of service*   *[350 000; +∞]*

Profitability of the motor vehicle fleet [kg/(m³ × year)] (only for utilities of retail systems):

- *Good quality of service*   *[450; +∞]*
- *Average quality of service*   *[400; 450]*
- *Unsatisfactory quality of service*   *[0; 400]*

Adequacy of human resources (No./1000 t):

For the utilities of bulk systems:

- *Good quality of service*   *[0.3; 0.6]*
- *Average quality of service*   *[0.2; 0.3] and [0.6; 0.7]*
- *Unsatisfactory quality of service*   *[0.0; 0.2] and [0.7; +∞]*
(for mainly urban areas served)
- *Good quality of service*   *[0.3; 0.65]*
- *Average quality of service*   *[0.2; 0.3] and [0.7; 0.8]*
- *Unsatisfactory quality of service*   *[0.0; 0.2] and [0.8; +∞]*
(for moderately urban areas served)
- *Good quality of service*   *[0.3; 0.8]*
- *Average quality of service*   *[0.2; 0.3] and [0.8; 0.9]*
- *Unsatisfactory quality of service*   *[0.0; 0.2] and [0.9; +∞]*
(for mainly rural areas served)

For the utilities of retail systems:

- *Good quality of service*   *[1.5; 2.5]*
- *Average quality of service*   *[1.0; 1.5] and [2.5; 3.0]*

- *Unsatisfactory quality of service*   *[0.0; 1.0] and [3.0; +∞]*

(for mainly urban areas served)
- *Good quality of service*   *[1.5; 3.0]*
- *Average quality of service*   *[1.0; 1.5] and [3.0; 3.5]*
- *Unsatisfactory quality of service*   *[0.0; 1.0] and [3.5; +∞]*

(for moderately urban areas served)
- *Good quality of service*   *[1.5; 3.5]*
- *Average quality of service*   *[1.0; 1.5] and [3.5; 4.0]*
- *Unsatisfactory quality of service*   *[0.0; 1.0] and [4.0; +∞]*

(for mainly rural areas served)

Use of energy resources (kWh/t for the utilities of bulk systems; tep/1000t for the utilities of retail systems):

For the utilities of bulk systems:
- *Good quality of service*   *[−∞; 6]*
- *Average quality of service*   *[6; 7]*
- *Unsatisfactory quality of service*   *[7; +∞]*

For the utilities of retail systems:
- *Good quality of service*   *[0; 6]*
- *Average quality of service*   *[6; 7]*
- *Unsatisfactory quality of service*   *[7; +∞]*

Quality of the leachate after treatment (%) (only for the utilities of bulk systems):
- *Good quality of service*   *[95; 100]*
- *Average quality of service*   *[75; 95]*
- *Unsatisfactory quality of service*   *[0; 75]*

Emission of greenhouse gases (kg CO_2/t):

For the utilities of bulk systems:
- *Good quality of service*   *[0; 50]*
- *Average quality of service*   *[50; 100]*
- *Unsatisfactory quality of service*   *[100; +∞]*

(for mainly urban areas served)
- *Good quality of service*   *[0;100]*
- *Average quality of service*   *[100; 200]*
- *Unsatisfactory quality of service*   *[200; + ∞]*

(for moderately urban areas served)
- *Good quality of service*   *[0; 200]*
- *Average quality of service*   *[200; 400]*
- *Unsatisfactory quality of service*   *[400; +∞]*

(for mainly rural areas served)

For the utilities of retail systems:
- *Good quality of service*   *[0; 13]*
- *Average quality of service*   *[13; 16]*
- *Unsatisfactory quality of service*   *[16; +∞]*

(for mainly urban areas served)
- *Good quality of service*   *[0; 14]*

- *Average quality of service*   *[14; 17]*
- *Unsatisfactory quality of service*   *[17; +∞]*
(for moderately urban areas served)
- *Good quality of service*   *[0; 15]*
- *Average quality of service*   *[15; 18]*
- *Unsatisfactory quality of service*   *[18; +∞]*
(for mainly rural areas served)

12.6 REGULATORY SYNERGIES

Quality of service regulation articulates closely with other components of the regulation model, to enable the corresponding synergies, as the information gathered here can and should be cross checked for validation, interpretation and analysis purposes, with information arising from the economic regulation, drinking water quality regulation and user interface regulation.

It may also lead to decisions regarding economic regulation, due to the possible need for new investments to improve quality of service, as well as contribute to better organisation of the sectors, through any need for strategic alterations, contribute to the clarification of rules of the sectors, through any need to alter legislation, and contribute towards capacity building for the sectors, through any need to carry out studies or training in this area.

Quality of service regulation is a way of regulating performances which are particularly inseparable from economic regulation, and constraining the permitted performances of the utilities regarding the services they provide to users.

This regulatory component also provides an important set of data for the regular production and disclosure of information concerning the sectors.

12.7 SUMMARY

This present chapter described in detail one of the components of the proposed regulatory approach, within the framework of behavioural regulation of the utilities, designated as quality of service regulation, including its respective goals, activities and procedures, instruments and synergies.

The next chapter will describe another of the proposed regulatory approach components, which also falls within the framework of behavioural regulation for the utilities, the drinking water quality regulation.

Chapter 13

Drinking water quality regulation

13.1 INTRODUCTORY NOTE

As part of this integrated approach (model RITA-ERSAR), this chapter describes drinking water quality regulation in more detail, one of the components of the regulation model for the public drinking water supply services to the population. This concerns a singly regulatory component which, for obvious reasons, is not applicable to waste water management and solid waste management services.

Although the content of this chapter is very much centred on drinking water quality regulation for the public supply of these services, it is important that this component of the regulatory model also envisages mechanisms to regulate private systems of supply, namely those that supply water to the public. This group includes, for example, hotels, restaurants and camping parks which, through their own sources, make drinking water available to their clients and patrons.

In those specific situations, the regulatory procedures to be applied should be adapted, although as a whole they should not be very different from those which are applied for public supply, with the primary consideration given to the need to protect public health.

13.2 REGULATORY GOALS

The public drinking water supply is the only one of the three services which supplies a product for food purposes, where quality control takes on an extraordinary level of importance. The quality of drinking water is an essential aspect of these services, particularly with regard to public health, and constitutes an element of great importance within the wider concern for quality of service. Reinforced by its characteristics as an irreplaceable and heterogeneous product located in space and time, the water supplied by the utilities should always be subject to suitable supervision in terms of quality, with this need reinforced by the fact that the service is provided as a natural monopoly.

This component of the regulatory model should therefore have the goal of ensuring in a continuous manner the supply of an appropriate level of drinking water quality by utilities, in accordance with applicable law, for the benefit of public health.

It will thus contribute to attaining public policy goals for the supply of water and meet public service requirements in terms of quality, as defined in 2.2 and in 3.3.

13.3 REGULATORY ACTIVITIES AND PROCEDURES

The regulator should ensure the existence of a suitable system to assess water quality, as realised through both legislation and regulations, and also any supplementary documentation. It must define the assessment criteria and the reference values or levels, specifically the quality goals to be attained, and in this way enable the utilities to define and carry out a suitable management strategy and also enable analysis laboratories to have the capacity to be able to work with the utilities in this activity.

It should annually carry out a cycle of water quality regulation based on that system, applicable to all the utilities providing public drinking water supply services, using a set of procedures.

In this cycle, which should take place in an on-going and planned manner, the regulator should follow a specific, clear and rational procedure, as laid down in a set of regulatory procedures. This regulatory activity should be carried out by each utility in three distinct periods of time, before the reference year (*ex-ante*), during the reference year and after the reference year (*ex-post*).

There follows an example of a regulatory procedure that the regulator can run. It is divided into the following stages:

In the period before the reference year (*ex-ante*):

- Each utility should submit for the approval of the regulator, using the water quality regulation module from the information system, its programme for the legal assessment of drinking water quality for the following year, drawn up in accordance with the terms defined in legislation, regulations and supplementary documentation, within a period before the start of the year to which it refers, indicating and identifying the laboratory responsible for carrying out the analyses. The laboratory carrying out the analyses should have previously submitted a suitable document to the regulator proving its scope of accreditation for analysing the parameters in accordance with national or international standardisation (e.g., accreditation in accordance with the ISO 1702 standard), issued by a competent accreditation body, and submit an updated version of this document whenever there have been alterations which have an effect on its accreditation period.
- The regulator should analyse and, where appropriate, approve the programmes for the legal assessment of drinking water quality by the final day of the year prior to that which they refer, including the acceptance of the laboratories specified, using the drinking water quality regulation module of the information system. Should it consider that the programmes do not meet the necessary conditions for approval, the regulator should return them to the utilities and establish a period for the latter to carry out any necessary modifications so that they can be approved. The programmes should be considered as having been tacitly approved if the regulator has not issued an opinion in their regard by that date, so as not to prejudice the assessment of the water quality.

  In the case of private supply systems, for instance hotels with their own water supply, and the generally large number of such entities, this procedure may be substituted by the simple requirement that the relevant entities keep updated records of the water quality assessment carried out in accordance with applicable legislation, which are open to inspection.

In the period corresponding to the reference year:

- Throughout the reference year the utilities should implement their approved programmes for the legal assessment of drinking water quality, carrying out all the envisaged analyses at the places and on the dates duly scheduled, informing the regulator of any situations of non-compliance with the threshold values, their respective causes, corrective measures and the results of analyses carried out to verify the latter, using the drinking water quality regulation module from the information system.

Taking into account the risk to public health, in the event of non-compliance with the threshold values the regulator should immediately monitor the resolution of such cases. The utilities should communicate these cases of non-compliance by the next working day following them becoming aware of such cases, as well as situations where such non-compliance remains, thus ensuring their timely resolution.

All data regarding the resolution of non-compliance cases which have been verified should be recorded in the information system for future recall and the systematising of such knowledge.

Communication of water quality assessment results by private supply systems may be substituted by the requirement that they disclose such results in their respective supply premises so as to provide information for users and also to record and store such information for the purposes of inspection.

- The regulator should carry out random inspections of the utilities to ensure compliance with the approved legal assessment programmes for drinking water quality as well as applicable legislation, based specifically on using the drinking water quality regulation module from the information system. In an initial stage the predominant factors for selecting the utilities to be monitored should be empirical knowledge of their performance and their compliance percentages with minimum regulatory frequency and the threshold values. At a more advanced stage, risk analysis tools should be utilised where the selected criteria are, for example, the non-compliance percentage with threshold values, the treatment carried out for threshold values' non-compliance, the presence of contractual problems, the date of the last inspection, the existence of complaints regarding water quality and the size and organization of the utilities.
- The regulator should also carry out inspections of the laboratories undertaking the analyses, in conjunction with the respective accreditation body, since the reliability of the results obtained by the laboratories is a determining factor in the implementation of the legal assessment programme for the water quality by the utilities.
- When any instance of legal non-compliance has been ascertained, the regulator should instigate administrative infringement proceedings against the utilities, in the manner set out in legislation, particularly using the information available in the drinking water quality regulation module of the information system.

In the period following the reference year (*ex-post*):

- By a pre-defined date after the end of the monitoring year the utilities should communicate to the regulator, using the drinking water quality regulation module of the information system, all the results obtained from verifying the water quality and implementing the legal assessment programme for drinking water quality, including results which were compliant and those which were non-compliant.
- The regulator should carry out the validation of the results sent by the utilities, and detect any data entry errors through cross analysis and validation and provision of any clarifications in the event of doubts, using the drinking water quality regulation module from the information system.
- After verification, the regulator should carry out the processing and interpretation of the water quality results sent by the utilities, and assess legal compliance with the carrying out of the analyses and compliance with the legal threshold values for water quality, using the drinking water quality regulation module of the information system.
- The regulator should establish a contradictory period, sending the utilities its assessment regarding compliance with the carrying out of the analyses and compliance with the legal threshold values for water quality, and setting aside a pre-defined period to submit comments or provide a suitably justified correction of the assessment, using the drinking water quality regulation module of the information system.

- Based on information which has been validated and responded to, the regulator should process the definitive data and interpret the results, with individual assessment of all the water quality parameters for each utility with semaphoric codes, with global assessment for groups of utilities per each parameter, or benchmarking, also with semaphoric codes, and the assessment of the evolution of the parameters by each utility, region and country, using the drinking water quality regulation module from its information system.

Figure 13.1 Drinking water quality regulation cycle.

- The regulator should publicly disclose the results of this regulatory component through an annual report on water services, for professional use, and through interactive apps, for non-professional use, on its website. It is indeed important to ensure the regular public disclosure of assessment results concerning water quality conformity and any instances of non-compliance which may have occurred, but in such a way that can be easily understood by everyone, particularly the users.

In addition, based on the experience acquired from the annual regulation cycle and from monitoring instances of non-compliance, the regulator should identify the need and if necessary work for the implementation of any methodological, procedural, legislative, regulatory or any other alterations to do with the application of this regulatory component to improve drinking water quality.

The different stages of the drinking water quality regulation cycle are shown in Figure 13.1.

13.4 REGULATORY INSTRUMENTS

The regulator must promote, develop and use the most appropriate instruments for the regulation of the quality of drinking water for each utility. Various examples of regulatory instruments will be presented below:

- *Water quality legislation*: To carry out this regulatory component, the regulator must base its work on national legislation regarding the quality of water for human consumption, which sets forth the rules to be obeyed by utilities in the supply of water to users, through a greater or lesser control of physical, chemical and bacteriological parameters, the risks from pathogens being by far the most important. If this does not exist, the regulator should be based on internationally recognised documentation serving as a guideline in this area, such as the guidelines of the World Health Organisation.

  Legislation in Portugal concerning the quality of drinking water is embodied in Decree-Law No. 306/2007, of 27 August, which transposes European legislation.

- *Water quality regulation*: In addition to legislation, the regulator should be guided by, should this exist, additional water quality rules or recommendations, giving a more detailed definition of the operational aspects which support the activities of the utilities.

  ERSAR has drawn up various recommendations on drinking water quality, as a supplement to existing legislation, and these are also available at its website (www.ersar.pt), mostly in Portuguese:

  ○ *Recommendation No. 2/2005: Assessment of lead in drinking water.*
  ○ *Recommendation No. 3/2005: Assessment of iron and manganese in drinking water.*
  ○ *Recommendation No. 4/2005: Assessment of arsenic in drinking water.*
  ○ *Recommendation No. 5/2005: Alternative method for the analysis of coliform bacteria and escherichia coli.*
  ○ *Recommendation No. 7/2005: Assessment of bromides in drinking water.*
  ○ *Recommendation No. 2/2006: Good practices in purchasing products utilised in treating drinking water.*
  ○ *Recommendation No. 5/2007: Disinfection of water used as drinking water.*
  ○ *Recommendation No. 1/2008: Communication and correction of non-compliance with parameter values for drinking water quality.*
  ○ *Recommendation No. 2/2008: Correction of the aggressiveness of water for human consumption in small population clusters.*
  ○ *Recommendation No. 3/2008: Assessment of the quality of drinking water in private supply systems.*
  ○ *Recommendation No. 3/2010: Procedure for collecting drinking water samples in supply systems.*

  ○ *Recommendation No. 2/2011: Technical specification for the certification of the drinking water product.*
  ○ *Recommendation No. 3/2011: Quarterly publication of data on drinking water quality.*
  ○ *Recommendation No. 4/2011: Assessment of risk when determining taste in drinking water samples.*

- *Programme for the legal assessment of water quality*: To carry out this regulatory component, the regulator should act in accordance with the programme for the legal assessment of water quality, which should include identifying the utility, identifying and localising the water sources, identifying and localising the supply areas, describing the treatment of the water, the annual average daily volumes entering the supply areas, the population serviced by the supply areas, the sampling schedule, the list of parameters to be analysed for each type of assessment and the laboratory used to carry out the analyses.

  In Portugal, ERSAR produced its Technical Guide 6, entitled 'Assessment of the quality of drinking water in public supply systems', published in 2005, and available at its website (www.ersar.pt), to provide support for the utilities in implementing a programme for the legal assessment of water quality.

- *Programme for the operational control of water quality*: In addition, the regulator should also act in accordance with the programme for the operational control of the water quality of the utility, which should ideally exist, describing the operating assessment which enables it to permanently distribute water at a suitable level of quality, minimising the risks to public health, complying with legislative provisions and reducing consumer complaints.

  In Portugal, ERSAR produced its Technical Guide 10, entitled 'Operational control in public supply systems', published in 2007, to provide support for the utilities in implementing a programme for the operational control of water quality, available at its website (www.ersar.pt). It also produced its Technical Guide 11, entitled 'Protecting surface and underground sources in public water supply systems', published in 2009, and Technical Guide 13, entitled 'The treatment of drinking water and the quality of source water', published in 2009.

- *Water safety plan*: Operational control constitutes a first step towards a desirable water safety plan, an instrument which has gradually been adopted by utilities seeking to detect and correct, in real time, any alterations which may occur regarding water quality. If such a situation exists, the regulator should also act based on the water safety plan, which should include risk assessment, operational planning, emergency planning, operational procedures, a contingency plan and a maintenance plan.

  The safety plan for drinking water quality, as described in the World Health Organisation guidelines, is a document which identifies credible risks from the abstraction of water to the user's tap, classifies these risks in terms of priority and establishes control mechanisms which can reduce them. The plan also requires processes to verify the effectiveness of the management control systems set up and the quality of the water produced. The safety plan is increasingly a regulatory requirement that can be applied equally in developed and developing countries, although the approaches and level of detail need to match the local circumstances.

  According to the Bonn Charter for Safe Drinking Water of the International Water Association there are three important stages which make up an efficient safety plan: the systematic assessment of risks from water abstraction to the user's tap, the identifying and monitoring of the most suitable control points to reduce identified risks and the development of efficient management systems and operational plans to deal with normal and exceptional operating conditions. Attention should be given to the potential occurrence of serious incidents and the management initiatives that should happen in such events.

Assessment of the efficiency of the management control systems is also essential, and the plan should include the carrying out of measures to assess the effectiveness of the controls which have been set up, and a check made on the efficiency of the control systems by an independent third party.

In addition to this, the management control systems should specify the responsibilities, documented procedures and a training plan to ensure that the respective human resources will have the necessary experience and suitability for the scale and complexity of the supply system. It might be more appropriate to use a simplified system for smaller supply systems.

In Portugal, ERSAR, in collaboration with the University of Minho, produced its Technical Guide 7, entitled 'Drinking water safety plans', published in 2005, to provide support for the utilities in implementing a safety plan for water, which is available at its website (www.ersar.pt).

- *Annual report*: The regulator should publish an up-to-date annual services report on the situation regarding drinking water quality, aimed at all stakeholders requiring reliable information, both to support the definition of policies and business strategies, as well as to assess the services actually provided to society.

In Portugal, ERSAR has published the annual report on water and waste services in Portugal (RASARP) since 2004. It includes a volume assessing drinking water quality, and is available at its website (www. ersar.pt).

- *Inspections and audits*: The regulator should carry out inspections whenever it sees fit to do so as part of its powers of authority, to ensure ongoing assessment of utility compliance with legislation. It should also carry out audits whenever it sees fit to do so as part of its powers of authority, to ensure the reliability of the information supplied by the utility. These inspections and audits should also include laboratories undertaking analyses, so as to verify the validity of water quality test results.

In Portugal, ERSAR supplies a standard inspection report, which covers the verification of all legal norms which may result in offences, the results of the meeting held with the utility and sometimes the health authority, as well as a set of recommendations aimed at improving the performance of the utility.

- *Disclosure of reference cases*: The regulator should include the annual awarding of prizes within the assessment system for drinking water quality, in order to identify, reward and share reference cases concerning the public drinking water supply services. Furthermore, this activity gives the opportunity to share specific cases where utilities constitute reference cases and to raise the awareness of utilities and the sectors in general. There are of course many possible forms of awarding such a distinction, and this may for example involve the annual awarding of seals of approval and prizes for excellence. The seals may be awarded to a greater or lesser number of utilities providing services which, in the previous annual regulatory assessment period, complied with a set of previously established and highly demanding criteria. In addition, prizes for excellence may be awarded to a utility which, in addition to meeting the criteria to be awarded a seal for drinking water quality, has shown outstanding performance or a remarkable improvement.

In Portugal, ERSAR promotes annually, together with a newspaper, some technical and scientific associations and a research center, the quality of service awards, including drinking water quality, recognizing the excellency of the nominated utilities.

- *Information system*: The regulator should make use of an information system to increase its regulatory effectiveness and efficiency, with regard to the high volume of information which its activity creates, along with a set of applications which enable access by its technicians and by the utilities. This should include a drinking water quality module.

  In Portugal, ERSAR provides a sophisticated information system, which is an indispensable instrument for its daily activity, which has various modules including one specifically for the assessment of drinking water quality.

  ERSAR also provides utilities with support for entering data relating to drinking water quality through its 'User guide for entering water quality data – IDQA', published in 2005.

- *Penalties system*: The regulator should adopt a suitable penalties system, to impose penalties on utilities for acts or omissions infringing legal provisions relating to drinking water quality.
- *Rules on regulatory procedures*: The regulator must have rules on regulatory procedures that define in detail the procedures concerning its relations with the utilities under its regulation, as part of the duties and competences invested in it by law, in terms of drinking water quality.

The regulatory activity described, undertaken with the above-mentioned regulatory procedures and these regulatory instruments, allows the regulator to effectively and efficiently achieve the goals for drinking water quality regulation, one of the components of the regulation model for public drinking water supply services to the population.

13.5 DRINKING WATER QUALITY ASSESSMENT SYSTEM

13.5.1 Overview

As far as regulation of the quality of drinking water by the regulator, as supplied by the utilities is concerned, and taking into account the complexity of the issue, use of a suitable assessment system becomes essential, as mentioned above. This assessment system enables data to evolve towards information and from this towards effective knowledge, which can particularly be used in regulatory terms.

The importance of this drinking water quality assessment system stems not only from being a powerful instrument to promote greater effectiveness and efficiency on the part of the utilities, but also as it promotes an improvement in the quality of water which is supplied, as well as embodying a basic right of the users of these services, namely that of having access to reliable and easily interpreted information on the water services being provided to them.

The drinking water quality assessment system is made up of the following components, which are the water quality indicators, the water sampling frequency and the reference values:

- The most important component of the system naturally consists of the water quality indicators, which enable a quantitative measurement to be determined for drinking water quality as supplied by the utilities and assess, in a quantified manner, compliance with the main goals laid down for this service, using reference values. They facilitate the assessment of compliance with goals and analysis of its evolution over time, thus simplifying an analysis which by its nature could be complex.
- The system needs to define the water sampling frequency, a function of the relevant indicators, to ensure the representativeness of that sampling and, as a result, the effectiveness of the assessment.
- The system also requires reference values, to be specified in legislation or by the regulator, which should reflect the limits considered desirable for each water quality indicator.

Assessment of the quality of the service is thus carried out based on the interpretation of the water quality indicators, comparing the values determined in the laboratory carrying out the analysis with the previously defined reference values, and attributing, for example, good, average or unsatisfactory performance, in accordance to its position within the reference values, in terms of being close or significantly removed from these.

Based on information which has been validated and responded to, the regulator should process the definitive data and produce:

- The individual assessment for each utility in terms of the water quality indicators selected and for the safe water aggregate indicator, with the performance analysis illustrated with a semaphoric code.
- The benchmarking of all the utilities, for each water quality indicator and for the safe water aggregate indicator, at the level of each utility, at the level of each group of utilities and at a national level, with the performance analysis illustrated with a semaphoric code.
- The historical evolution of each water quality indicator and the aggregate indicator, as regards each utility, each group of utilities and at national level.

Figure 13.2 shows the components of the drinking water quality assessment system and the flow of data which occurs, which is described in greater detail below:

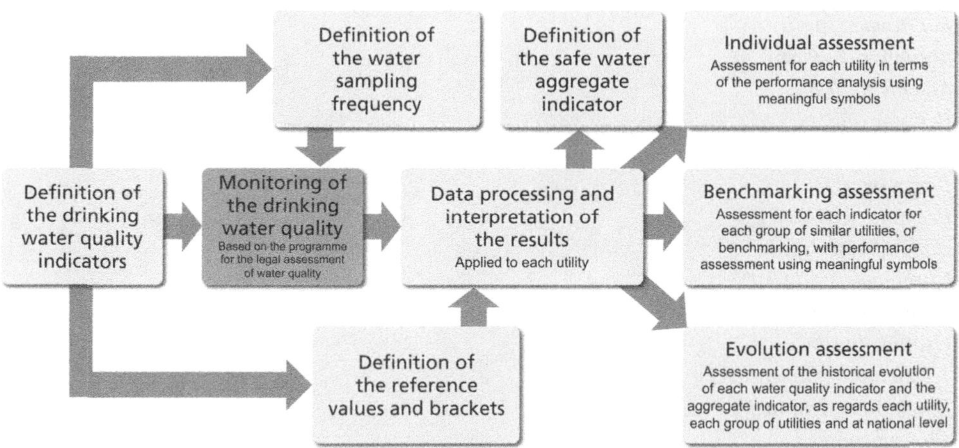

Figure 13.2 Drinking water quality assessment system.

Each of the components of the system to assess the quality of drinking water will now be presented in greater detail.

13.5.2 Water quality indicators

Legislation or, in its absence, rules of the regulator should specify the drinking water quality indicators, that is, the aspects which it considers may jeopardise public health and user acceptability of the water, and even long term protection from the materials in contact with, based on the best available scientific knowledge and on consumer demands.

The process of defining the quality standards should be technically sound and transparent, disclosing the choices made, taking into account the level of risk considered acceptable to public health.

The choices of each country are often based on internationally recognised documentation to serve as a guideline in this issue, such as the guidelines of the World Health Organisation or the Directives of the European Commission.

Although setting health-based standards is desirable at the theoretical point of view, it is complex to do in developing countries and almost impossible in developing countries, due to the lack of epidemiologic information and the complexity of the correlation between water quality and epidemiology. That means health-based approach is not necessary to justify investment in water and waste services and hygiene, also because their benefits greatly exceed the costs.

In developing countries, where it may take a long time and very important investments to achieve standards based on World Health Organisation guidelines, it is preferable to define a first set of interim targets based on protection from pathogens alone, unless there are specific concerns over chemicals at toxic contaminants.

In Portugal, ERSAR uses the following drinking water quality indicators, which total more than fifty, in conformity with national legislation which results from the transposition of European legislation:

Obligatory parameters (microbiological and chemical)

- *Escherichia coli (No./100 ml)*
- *Enterococci (No./100 ml)*
- *Acrylamide (µg/l)*
- *Antimony (µg/l Sb)*
- *Arsenic (µg/l As)*
- *Benzene (µg/l)*
- *Benzo(a)pyrene (µg/l)*
- *Boron (mg/l B)*
- *Bromate (µg/l BrO₃)*
- *Cadmium (µg/l Cd)*
- *Chromium (µg/l Cr)*
- *Copper (mg/l Cu)*
- *Cyanide (µg/l Cn)*
- *1,2 dichloroethane (µg/l)*
- *Epichlorohydrin (µg/l)*
- *Fluoride (mg/l F)*
- *Lead (µg/l Pb)*
- *Mercury (µg/l Hg)*
- *Nickel (µg/l Ni)*
- *Nitrate (mg/l NO₃)*
- *Nitrite (mg/l NO₂)*
- *Individual pesticides (µg/l)*
- *Pesticides – total (µg/l)*
- *Polycyclic aromatic hydrocarbons (HAP) (µg/l)*
- *Selenium (µg/l Se)*
- *Tetrachloroethene and trichloroethene (µg/l)*
- *Trihalomethanes – total (THM) (µg/l)*
- *Vinyl chloride (µg/l)*

Parameter indicators (microbiological, physical, chemical and organoleptic)

- *Aluminium (μg/l Al)*
- *Ammonium (mg/l NH_4 Calcium - mg/l Ca)*
- *Chloride (mg/l Cl)*
- *Clostridium perfringens (including spores) (No./100 ml)*
- *Colour (mg/l PtCo)*
- *Conductivity (μS/cm at 20°C)*
- *Total hardness (mg/l $CaCO_3$)*
- *pH (pH units)*
- *Iron (μg/l Fe)*
- *Magnesium (mg/l Mg)*
- *Manganese (μg/l Mn)*
- *Microcystins - LR total (μg/l)*
- *Smell, at 25°C (dilution factor)*
- *Oxidisability (mg/l O_2)*
- *Sulphates (mg/l SO_4)*
- *Sodium (mg/l Na)*
- *Taste, at 25°C (dilution factor)*
- *Number of colonies (No./ml at 22°C)*
- *Number of colonies (No./ml at 37°C)*
- *Coliform bacteria (No./100 ml)*
- *Total organic carbon (TOC) (mg/l C)*
- *Turbidity (UNT)*
- *α-total (Bq/l)*
- *β-total (Bq/l)*
- *Tritium (Bq/l)*
- *Total indicative dose (mSv/year)*
- *Residual disinfectant (mg/l)*

13.5.3 Water sampling frequency

Legislation or, in its absence, rules from the regulator should set forth the minimum sampling frequency and drinking water analysis for each indicator, covering water distributed not only through a distribution network but also by fountains, by a tanker lorry or by food industry enterprises.

The minimum sampling frequency should be what is necessary to ensure the representativeness of the samples, that is to say closely translating what actually happens in reality.

In Portugal, ERSAR uses an annual sampling frequency for drinking water, as a function of the volume of water supplied (cubic metres per day) in each supply zone, in conformity with national legislation which results from the transposition of European legislation and which varies according to whether this involves an inspection or a routine assessment.

Both have the aim of demonstrating conformity with the reference values, but the inspection control is a global assessment, that is to say it seeks to ascertain if there is conformity with all reference values, including the routine control elements. The aim of the routine assessment is merely to provide more information on issues related to acceptability and the parameters that could have a more immediate impact on the protection of human health (microbiological) and on treatment processes. The routine assessment is intended to undertake more regular verification, with a greater sampling frequency.

The following annual sampling frequency is adopted in the routine assessment:

Routine assessment	Volume of water supplied	Annual sampling frequency
• *Escherichia coli (E. coli)*	*<100 m³/day*	*6*
• *Coliform bacteria*	*≥100 m³/day*	*12/5 000 inhab.*
• *Residual disinfectant*	*≥100 m³/day*	*12/5 000 inhab.*
• *Aluminium*	*<100 m³/day*	*2*
• *Ammonium*	*<100 m³/day*	*2*
• *Number of colonies at 22°C*	*<100 m³/day*	*2*
• *Number of colonies at 37°C*	*<100 m³/day*	*2*
• *Conductivity*	*>100 and ≤1 000*	*4*
• *Clostridium perfringens*	*>100 and ≤1 000*	*4*
• *Colour*	*>100 and ≤1 000 m³/day*	*4*
• *pH*	*>1 000*	*4 + 3 per 1000 m³/day + 3 per remaining fraction*
• *Iron*	*>1 000*	*4 + 3 per 1000 m³/day + 3 per remaining fraction*
• *Manganese*	*>1 000*	*4 + 3 per 1000 m³/day + 3 per remaining fraction*
• *Nitrate*	*>1 000*	*4 + 3 per 1000 m³/day + 3 per remaining fraction*
• *Nitrite*	*>1 000*	*4 + 3 per 1000 m³/day + 3 per remaining fraction*
• *Oxidisability*	*>1 000*	*4 + 3 per 1000 m³/day + 3 per remaining fraction*
• *Smell*	*>1 000*	*4 + 3 per 1000 m³/day + 3 per remaining fraction*
• *Taste*	*>1 000*	*4 + 3 per 1000 m³/day + 3 per remaining fraction*
• *Turbidity*	*>1 000*	*4 + 3 per 1000 m³/day + 3 per remaining fraction*

In the assessment inspection the following annual sampling frequency is adopted:

Assessment inspection	Volume of water supplied	Annual sampling frequency
• *Antimony*	*≤1 000*	*1*
• *Arsenic*	*≤1 000*	*1*
• *Benzene*	*≤1 000*	*1*
• *Benzo(a)pyrene*	*>1 000 and ≤10 000*	*1 + 1 per 3 300 m³/day + 1 per remaining* fraction
• *Boron*	*>1 000 and ≤10 000*	*1 + 1 per 3 300 m³/day + 1* per remaining fraction
• *Bromate*	*>1 000 and ≤10 000*	*1 + 1 per 3 300 m³/day + 1 per remaining* fraction
• *Cadmium*	*>1 000 and ≤10 000*	*1 + 1 per 3 300 m³/day + 1 per remaining* fraction
• *Calcium*	*>1 000 and ≤10 000*	*1 + 1 per 3 300 m³/day + 1 per remaining* fraction
• *Lead*	*>1 000 and ≤10 000*	*1 + 1 per 3 300 m³/day + 1 per remaining* fraction
• *Cyanide*	*>1 000 and ≤10 000*	*1 + 1 per 3 300 m³/day + 1 per remaining* fraction
• *Copper*	*>10 000 and ≤100 000*	*3 + 1 per 10 000 m³/day + 1 per remaining* fraction
• *Chromium*	*>10 000 and ≤100 000*	*3 + 1 per 10 000 m³/day + 1 per remaining* fraction
• *1,2 dichloroethane*	*>10 000 and ≤100 000*	*3 + 1 per 10 000 m³/day + 1 per remaining* fraction
• *Total hardness*	*>10 000 and ≤100 000*	*3 + 1 per 10 000 m³/day + 1 per remaining* fraction
• *Enterococci*	*>10 000 and ≤100 000*	*3 + 1 per 10 000 m³/day + 1 per remaining* fraction
• *Fluoride*	*>100 000*	*10 + 1 per 25 000 m³/day and remaining* fraction

- *Magnesium*   *>100 000*   *10 + 1 per 25 000 m³/day and remaining* fraction
- *Mercury*   *>100 000*   *10 + 1 per 25 000 m³/day and remaining* fraction
- *Nickel*   *>100 000*   *10 + 1 per 25 000 m³/day and remaining* fraction
- *HAP*   *>100 000*   *10 + 1 per 25 000 m³/day and remaining* fraction
- *Individual Pesticides*   *>100 000*   *10 + 1 per 25 000 m³/day and remaining* fraction
- *Pesticides (total)*   *>100 000*   *10 + 1 per 25 000 m³/day and remaining* fraction
- *Selenium*   *>100 000*   *10 + 1 per 25 000 m³/day and remaining* fraction
- *Chloride*   *>100 000*   *10 + 1 per 25 000 m³/day and remaining* fraction
- *Tetrachloroethene and trichloroethene*   *>100 000*   *10 + 1 per 25 000 m³/day and remaining* fraction
- *Trihalomethanes – Total*   *>100 000*   *10 + 1 per 25 000 m³/day and remaining* fraction
- *Sodium*   *>100 000*   *10 + 1 per 25 000 m³/day and remaining* fraction
- *Total organic carbon*   *>100 000*   *10 + 1 per 25 000 m³/day and remaining* fraction
- *Sulphate*   *>100 000*   *10 + 1 per 25 000 m³/day and remaining* fraction
- *Vinyl chloride*   *>100 000*   *10 + 1 per 25 000 m³/day and remaining* fraction
- *Epichlorohydrin*   *>100 000*   *10 + 1 per 25 000 m³/day and remaining* fraction
- *Acrylamide*   *>100 000*   *10 + 1 per 25 000 m³/day and remaining* fraction

13.5.4 Reference values

Legislation or, in its absence, rules from the regulator should specify reference values for indicators of drinking water quality which should reflect the limits considered to be desirable.

In Portugal, ERSAR uses the following reference values for drinking water quality indicators, in conformity with national legislation which results from the transposition of European legislation:

Microbiological parameters	*Reference Values*
- *Escherichia coli (E.coli)*	*No./100 ml*
- *Enterococci*	*No./100 ml*

Chemical parameters	
- *Acrylamide*	*0.10 µg/l*
- *Antimony*	*5.0 µg/l Sb*
- *Arsenic*	*10 µg/l As*
- *Benzene*	*1.0 µg/l*
- *Benzo(a)pyrene*	*0.010 µg/l*
- *Boron*	*1.0 mg/l B*
- *Bromate*	*10 µg/l BrO₃*
- *Cadmium*	*5.0 µg/l Cd*
- *Chromium*	*50 µg/l Cr*
- *Copper*	*2.0 mg/l Cu*
- *Cyanide*	*50 µg/l Cn*
- *1,2 dichloroethane*	*3.0 µg/l*

- *Epichlorohydrin* — *0.10 µg/l*
- *Fluoride* — *1.5 mg/l F*
- *Lead* — *10 µg/l Pb*
- *Mercury* — *1 µg/l Hg*
- *Nickel* — *20 µg/l Ni*
- *Nitrate* — *50 mg/l NO$_3$*
- *Nitrite* — *0.5 mg/l NO$_2$*
- *Individual Pesticides* — *0.10 µg/l*
- *Pesticides – total* — *0.50 µg/l*
- *Polycyclic Aromatic Hydrocarbons (HAP)* — *0.10 µg/l*
- *Selenium* — *10 µg/l Sand*
- *Tetrachloroethene and trichloroethene* — *10 µg/l*
- *Trihalomethanes – total (THM)* — *100 µg/l*
- *Vinyl chloride* — *0.50 µg/l*

Parameter indicators (microbiological, physical, chemical and organoleptic)

- *Aluminium* — *200 µg/l Al*
- *Ammonium* — *0.50 mg/l NH$_4$ Calcium - mg/l Ca*
- *Chloride* — *250 mg/l Cl*
- *Clostridium perfringens (including spores)* — *No./100 ml*
- *Colour* — *20 mg/l PtCo*
- *Conductivity* — *2 500 µS/cm at 20°C*
- *Total hardness* — *– mg/l CaCO$_3$*
- *pH* — *≥ 6.5 and ≤9 pH units*
- *Iron* — *200 µg/l Fe*
- *Magnesium* — *– mg/l Mg*
- *Manganese* — *50 µg/l Mn*
- *Microcystins – LR total* — *1 µg/l*
- *Smell, at 25°C* — *Dilution factor 3*
- *Oxidisability* — *5 mg/l O$_2$*
- *Sulphates* — *250 mg/l SO$_4$*
- *Sodium* — *200 mg/l Na*
- *Taste, at 25°C* — *Dilution factor 3*
- *Number of colonies* — *No abnormal change No./ml at 22°C*
- *Number of colonies* — *No abnormal change No./ml at 37°C*
- *Coliform bacteria* — *No./100 ml*
- *Total organic carbon (COT)* — *No abnormal change mg/l C*
- *Turbidity* — *4 UNT*
- *α-total* — *0.5 Bq/l*
- *β-total* — *1 Bq/l*
- *Tritium* — *100 Bq/l*
- *Total indicative dose* — *0.10 mSv/year*
- *Residual disinfectant* — *–mg/l*

13.5.5 Water quality aggregate indicator

It is possible and desirable to define an aggregate indicator for water quality, designated indicator for safe water, covering water quality assessment as a whole, for example applicable at the utility, regional and national level.

This indicator could be defined as the percentage of controlled and good quality water, that is to say the product of the sampling frequency compliance percentage by the compliance percentage of the reference values established in legislation, for the indicators subject to quality control.

In Portugal, ERSAR has adopted and frequently uses the safe water indicator, defined as the percentage of all the analyses carried out as required that satisfy the parameter values, and this is calculated annually on a national and regional basis, as well as for each utility. It uses the following reference values for this indicator, in percentage terms: good quality of service [98.50; 100.00]; average quality of service [94.50; 98.50]; unsatisfactory quality of service [00.00; 94.50].

13.6 REGULATORY SYNERGIES

The drinking water quality regulation articulates closely with other components of the regulation model, to enable the corresponding synergies, since the information gathered here can and should be cross checked for validation, interpretation and analysis purposes with information arising from the quality of service and the user interface regulation.

It may also lead to decisions regarding economic regulation, owing to the possible need for new investments to improve drinking water quality, as well as the contribution to better organisation of the sectors, the possible need for strategic alterations, clarification of the rules of the sectors, the possible need to alter legislation, the capacity building of the sectors, any need for innovation or the conducting of studies or training in this area.

This regulatory component also provides an important set of data for the regular production and disclosure of information concerning the sectors.

13.7 SUMMARY

This chapter provided a detailed description of one of the components of the proposed regulatory approach, within the framework of behavioural regulation for utilities, known as drinking water quality regulation, including its respective goals, activities and procedures, instruments and synergies.

The following chapter will describe another of the components of the proposed regulatory approach, which also forms part of the behavioural regulation framework for utilities, the user interface regulation.

Chapter 14

User interface regulation

14.1 INTRODUCTORY NOTE

As part of the integrated approach (model RITA-ERSAR) being proposed, this chapter describes in more detail the user interface regulation, one of the components of the regulation model for the public services of drinking water supply, waste water management and solid waste management.

14.2 REGULATORY GOALS

The goals of this regulatory model component include ensuring the protection of user rights, through compliance by the utilities with legislation for user protection, defence of the right to information and to make complaints and, subsequently, to improve the quality of the relationship of the utilities with its users.

Indeed, without prejudice to the information which must be reported by the utilities for regulatory purposes, the regulator should ensure that key information is made available to users concerning service provision, particularly regarding service provision regulations, tariffs in force, contractual conditions in force, drinking water quality control results, quality of service provided to users, service interruptions and contacts and public reception hours.

While complaints cannot be an objective indicator of the quality of the service provided, in so far as not all of them turn out to be justified, they constitute an important indicator of the perception that users have of the service, and further reflect a set of factors referring to the users, in particular their level of requirements and their demands with regard to service.

This component of the regulatory model thus helps to fulfil public service obligations, as defined in section 2.2, through the adoption of rules concerning good practices such as transparency, user participation and conflict resolution mechanisms.

It also contributes to fulfilling the public policy goals defined in section 3.3, ensuring user protection, awareness and participation, which is especially important in such services which constitute natural or legal monopolies.

14.3 REGULATORY ACTIVITIES AND PROCEDURES

The regulator should promote the establishment of clear rules regarding the interface between users and utilities, particularly the rights and responsibilities of each party, as embodied in legislation, in the service regulations of the utility and in the regulation of commercial relations.

It should audit the utilities in a selective manner based on a risk analysis, to ensure compliance with legal and contractual provisions regarding its interface with users, particularly the existence, suitability and clarity of service regulation, the suitability and clarity of user contracts, the suitability and clarity of the invoicing of services and the availability of multiple contacts for the users with the utilities.

It should monitor and support the resolution of complaints by users which have been registered in the complaints books of the utilities and also those submitted directly to it by users or by any other means, or those which have been resubmitted by other bodies, such as user protection associations. To do this it should adopt a clear, specific and rational procedure engaged in a regulatory procedural form, which allows for efficient monitoring of each utility in this area. There follows a possible example of a regulatory procedure that can be adopted by the regulator in resolving complaints. It is divided into a number of stages:

- Whenever users wish to do so, they may register a complaint in the complaints book of the utility. In the event that the complaint is made by the user directly to the regulator, the latter should forward it to the utility.
- The utility should forward any complaints to the regulator within a predefined period, accompanied by the respective analyses and the clarifications provided to the claimants, using the user interface regulation module of the information system.
- After receiving and carrying out an initial analysis of these items, and where it is considered necessary, the regulator may request any clarification needed to understand the situation from the utility or from the claimant, which should be submitted by these within a pre-defined period.
- Based on the factors and the arguments presented by the parties, the regulator should carry out any analysis of the complaints, informing the claimant and the utility of its preliminary conclusions, and opening a contradictory period for the parties with a pre-defined period.
- The regulator may reanalyse the complaint after the contradictory period, based on the replies received, informing the claimant and the utility of its definitive evaluation, within a pre-defined global period which naturally excludes the clarification and response periods, using the user interface regulation module of the information system.
- Whenever the regulator decides to alter the procedure or the decision initially taken by the utility, it should confirm the execution of this alteration to the utility within a pre-defined period, using the user interface regulation module of the information system.

If justified, the regulator may open administrative infringement proceedings against the utility, under the terms set out in legislation.

The regulator should publicly disclose the results of this regulatory activity, particularly through the water and waste services annual report, both with regard to audits on utilities to ensure the existence and suitability of legal and contractual instruments, as well as in terms of the resolution of complaints submitted by users.

The different stages of the user interface regulation cycle are shown in Figure 14.1.

At the same time, the regulator should carry out user opinion studies through regular polls, since these allow it to assess the perception of users with regard to service aspects such as prices, quality of service, water quality or public interface. Given that this information is known by the regulator through the carrying out of various regulatory components, it is important to compare the results of its monitoring with user perception and assess the need for any awareness campaigns aimed at the latter when it finds a clear disparity between consumer perception and its monitoring results.

The regulator should frequently carry out user awareness campaigns, for example through the use of leaflets and media, which can explain various important aspects to users in a simple and clear manner.

Figure 14.1 User interface regulation cycle.

The regulator should periodically reassess and identify the need for the resolution of any malfunctions based on the experience acquired with the user interface regulation and, where necessary, promote the implementation of amendments to legislation, regulations or other instruments in order to improve the operation of the sectors and the activity of the utilities.

14.4 REGULATORY INSTRUMENTS

The regulator should promote, develop and use the most appropriate instruments for the user interface regulation relative to each utility. There follow some examples of regulatory instruments:

- *User protection legislation*: The regulator must naturally follow user protection legislation for this regulatory component. This defines mechanisms for protecting users of these services that take the form of legal obligations especially applicable to these service providers. Its content can, for example, include making information available, prohibition of charging excessive prices, obligation to give prior notice of suspension due to non-payment, frequency of billing, minimum payment time and prescription and limitation periods for debts.

In Portugal user protection legislation is currently provided for in Law No. 23/96, of 12 July, altered by Laws No. 12/2008, of 26 February, and No. 24/2008, of 2 June.

- *Utility contract*: The regulator must naturally follow the existing contract between the holder and the utility to carry out this regulatory component, which may be a delegated management contract between the service holder, as the delegating entity, and the utility as the delegatee, or a concession between the service holder, as grantor, and the utility, as concessionaire, which define the relationship rules between the two parties.
- *Commercial relations regulation*: In addition to the general user protection legislation, the regulator should follow the regulations on commercial relations that set out the duties and obligations of the two parties, namely the utility and the user.
- *Audits:* The regulator should carry out audits whenever it sees fit to do so as part of its powers of authority, to ensure ongoing assessment of utility compliance with legislation.
- *Conciliation processes*: The regulator should mediate and reconcile conflicts involving the utilities, by analysing these, and promote the use of conciliation and arbitration between the parties as a means of resolving conflicts and taking the measures which it considers to be urgent and necessary.
- *Awareness campaigns*: The regulator should frequently carry out user awareness campaigns, for example through the use of leaflets or media, which can explain various important aspects to users in a simple and clear manner.

In Portugal, in recent years ERSAR has published a number of leaflets on subjects such as 'The rights and duties of consumers', 'The quality of tap water' and 'Domestic water metering systems' in the 'Awareness booklets' series. They are available on its website (www.ersar.pt).

- *Rules on regulatory procedures*: The regulator must have rules on regulatory procedures that define in detail the procedures concerning relations with the utilities under its regulation as part of the duties and powers invested in it by law, particularly in terms of user interface regulation.
- *Annual report:* The regulator should publish an up-to-date annual services report on the user interface regulation, aimed at all stakeholders requiring reliable information, both to support and define policies and business strategies, as well as to assess the services actually provided to society.

In Portugal, ERSAR has published the annual report on water and waste services in Portugal (RASARP) since 2004. It includes a volume assessing interface regulation, which is available at its website (www.ersar.pt).

- *Information system*: In view of the large amount of information that this activity generates, regulatory effectiveness and efficiency can clearly be improved by using an information system with a user interface module.

In Portugal, ERSAR provides a sophisticated information system, which is an indispensable instrument for its daily activity, which has various modules including one specifically for user interface regulation.

- *Penalties system*: There should be a penalties system allowing the regulator to impose penalties on utilities for acts or omissions infringing legal or contractual provisions regarding user interface.

The regulatory activity described, undertaken with the above-mentioned regulatory procedures and these regulatory instruments, allows the regulator to effectively and efficiently achieve the goals of its user interface with the users, one of the components of the regulation model for public drinking water supply services, waste water management and solid waste management.

14.5 REGULATORY SYNERGIES

The user interface regulation is closely connected to other components of the regulation model, as the information gathered here can and should be cross checked for validation, interpretation and analysis purposes, with information arising from economic regulation, the quality of service regulation and the water quality regulation.

It may also lead to decisions concerning intervention in the regulation of legal and contractual compliance, for example in carrying out inspections and audits to get to the bottom of the issues identified by the regulator or the users, and in economic regulation, such as the need for new investments to improve any aspect of the quality of service subject to systematic complaint. It may also lead to decisions regarding intervention to clarify rules of the sectors, to a need to alter legislation, the capacity building and innovation of the sectors, and the possible need to carry out studies or training in this area.

This component also provides an important set of data for the regular production and disclosure of information concerning the sectors.

14.6 SUMMARY

This chapter described in detail one of the components of the proposed regulatory approach, which also forms part of the behavioural regulation framework for the utilities, designated as the user interface regulation, including its respective goals, activities and procedures, instruments and synergies. This chapter concludes the description of all the components of the proposed regulatory approach.

The following chapter will consider the articulation of the regulator with the main stakeholders in the public drinking water supply, waste water management and solid waste management services.

Chapter 15

Articulation with third party entities

15.1 INTRODUCTORY NOTE

As part of this integrated approach (model RITA-ERSAR), this chapter will consider the necessary articulation of the regulator with the main stakeholders in the public drinking water supply, waste water management and solid waste management services.

15.2 ARTICULATION WITH THE SECTORS' STAKEHOLDERS

It is important to ensure the suitable articulation of the regulator with the main stakeholders, with the aim of promoting greater effectiveness regarding regulation, but also stakeholder awareness, achieving synergies, the transparency of procedures, the additional collection of information and also greater legitimisation of regulation.

A well-thought out articulation strategy should thus be developed by the regulator with:

- Government and parliament, holders of political power and those responsible for public policy definition and the approval of legislation.
- Public administration, particularly in terms of clarifying its competencies to lead and coordinate the policy of water and waste services, as well as the articulation of activities in the boundary areas of their respective mandates.
- The relationship with the utilities which are subject to regulation, which should be based on the principles of cooperation, loyalty and transparency, with mutual gains which are of benefit to users and society.
- The users and users associations, the recipients of these essential public services, particularly through the participation and resolution of doubts and potential complaints.
- Associations representing economic activities, which are also recipients of these essential public services.
- Non-governmental environmental protection organisations, with regard to the potential environmental impacts of these services.
- The most relevant technical and scientific institutions and associations in the sectors, potential partners for capacity building and innovation, particularly in terms of cooperation, collaboration

and association, within the scope of their duties, as embodied in studies, training, audits and joint publications.
- Peer national and international regulatory bodies, when this is shown to be necessary or useful in carrying out its respective duties, particularly the exchange of regulatory experiences.

The regulator thus manages to enhance its activity through its series of efforts, complementarity of responsibilities and the sharing of knowledge with all these entities.

As mentioned above, the provision of water and waste services, besides involving the intervention of the regulator, is also subject to the intervention of different public bodies, as this involves activities with an impact in areas which are the responsibility of environmental, water resources, waste management, and public health authorities (supply of drinking water), consumer protection (provision of essential public services) and competition (public markets).

The correct functioning of the sectors thus has to include suitable articulation between the different public bodies involved, through the clear definition of their respective responsibilities, so as to avoid any overlaps or omissions. It should also enable synergies to be obtained between the different entities and the reconciliation of the goals carried out by each one of them, at the least cost possible for the provision of water and waste services and for society.

Due to its importance in ensuring a suitable institutional framework, with a clear assignment of responsibilities for the public entities involved, the articulation of the regulator for water and waste services with the environmental, water resources, waste management, public health, consumer protection and competition authorities will now be considered.

15.3 ARTICULATION WITH THE ENVIRONMENTAL AUTHORITY

The environmental authority, which may include the water and the waste resources authority, is generally responsible for proposing, developing and monitoring the integrated and involved management of environmental policies and sustainable development, in a manner articulated with other sectoral policies and in collaboration with public and private entities which contribute for the same purpose. It must take into account the protection and recovery of the environment and the provision of high quality services to the citizens.

The relationship of the water and waste services regulator with the environmental authority should include:

- Articulation in the implementation of public policies for water and waste services with environmental public policies.
- Articulation in the preparation of legislation and regulation for the water and waste services with environmental legislation and regulations.
- Articulation in the collection, validation, processing and dissemination of water and waste services information with environmental information.
- Articulation in boundary zones concerning soil discharges, such as the final destination of sludge originating from supply water and waste water treatment systems.
- Articulation in boundary zones concerning emissions to the atmosphere, such as those originating from supply water, waste water and solid waste treatment.
- Reporting by the environmental authority concerning inspection activities carried out on water and waste service utilities, which are important for the regulator's activity, and, vice versa, the latter reporting situations which may indicate the presence of environmental infringements.

The responsibilities of those two entities, the water services regulator and the environmental authority, may and should therefore be distinct but complementary, without the risk of overlaps or omissions.

15.4 ARTICULATION WITH THE WATER RESOURCES AUTHORITY

The link between the water services and the water resources is of course very strong. Indeed, that articulation should be focused on both the use of water resources as raw material for the production of drinking water as well as the use of water resources as a receiving body for waste water.

Good institutional articulation between the entity responsible for the management of water resources and the water services regulator is therefore essential. This presupposes a clear identification of boundary zones between the activities of those bodies, as well as articulation in developing structured instruments, implementing public policies, interpreting legislation, regulations and contracts and also with regard to information management.

As regards the need for a clear identification of boundary zones, the interventions of the water resources authority are focused on those resources and form the interface with the water services, as they do with other water users, such as agriculture, industry, energy production, transportation, tourism and leisure. On the other hand, the intervention of the regulator is distinct, focusing on the regulation of water supply and waste water services, involving suppliers and users, and concerning itself with the sustainability of those services from an integrated perspective and within a context of effectiveness and efficiency. It therefore regulates just one of the various water user sectors. The boundaries of their different interventions are legally clear and embodied mainly through two intersecting points, namely water abstraction and the discharge of waste water, activities subject to licensing by the respective environmental authority.

As for articulation in the development of structuring instruments, such as the setting up of policies, approval of legislation or even the definition of specific procedures, these should ideally be subject to joint analysis by the different bodies before their approval, in order to obtain optimum equilibrium from the distinct perspectives. For example, the drawing up or revision of water resources legislation requires not only suitable assessment of its environmental effectiveness but also an economic impact assessment of the water services sectors, the stakeholders involved and finally the users, through cost-benefit studies. It is therefore extremely important for there to be prior consultation between the parties, whenever the drawing up of policies, the approval of legislation or even specific procedures in interface areas between water resources and water services is concerned.

As for articulation in terms of interpreting legislation, regulations and contracts, the public administration should not, for obvious reasons, have different interpretations on the same matter. It is important for this reason to identify the boundary zones, primarily associated with issues involving abstraction and discharge of water resources, in order to ensure uniformity both in the interpretation carried out by the various entities and also in the application of legislative instruments. Examples of this are the use of water resources by entities providing water services, licences for private abstractions in the case of public water services availability, the desalination of sea water, waste water discharges and the production and use of treated waste water.

As regards articulation concerning the management of information, there is, on the one hand, the need to avoid overlapping information requests to the utilities by different entities and, on the other hand, the need to ensure mutual access to information systems. The public administration must not collect the same information or similar information through different bodies, at different moments using different procedures. It is the duty of the water resources authority to carry out an inventory and maintain a record of the public water domain and set up and keep up-to-date information and water resource management systems, as well as promote communication and ensure the dissemination of information to ensure knowledge of water resources in terms of catchment areas. The regulator has the duty to coordinate and carry out the collection and dissemination of information regarding the sectors for public water supply, waste water and the respective utilities.

The water resources authority, which may be integrated in the environmental authority, is generally responsible for the development and application of national policies in the area of water resources and the general coordination, planning and licensing in the management of water resources, with a view to ensuring in particular its protection and planning. As such, it oversees the preservation of water resources and ensures their rational use.

The relationship of the regulator with the water resources authority should include:

- Articulation of the implementation of public policies for water services with public policies for water resources.
- Articulation of the preparation of legislation and regulation for the water services with water resources legislation and regulations.
- Articulation of the collection, validation, processing and dissemination of water services information with water resources information.
- Articulation for boundary aspects concerning the management of water sources aimed at the production of drinking water and all associated information, for example, geographical location, protection of sources, monitoring and licensing of use.
- Articulation of boundary zones concerning the abstraction of surface, underground and even coastal water resources, for the purposes of supplying water for human consumption.
- Articulation of boundary zones concerning discharges of waste water into surface, underground and coastal water resources, or into the soil.
- Reporting by the water resources authority concerning inspection activities carried out on water service utilities, which are important for the regulator's activity, and, vice versa, namely the latter reporting situations which may indicate the presence of infringements.

The responsibilities of those two entities, the water services regulator and the water resources authority, may and should therefore be distinct but complementary, without the risk of overlaps or omissions.

15.5 ARTICULATION WITH THE WASTE AUTHORITY

The waste authority, which may be integrated within the environmental authority, has general responsibility for the development and application of national policies in the area of waste and the general coordination, planning and licensing in waste management, with a view to ensuring in particular its organization and planning.

The relationship of the water and waste services regulator with the waste authority should include:

- Articulation in the implementation of public policies for solid waste services with more wide-ranging public policies for waste in general.
- Articulation in the preparation of legislation and rules for waste services with waste legislation and regulations.
- Articulation in the collection, validation, processing and dissemination of waste services information with waste information.
- Articulation in boundary zones concerning the recovery of solid waste, such as the production of recycled materials, compost, electricity or heat.
- Articulation in boundary zones concerning solid waste disposal, such as channelling part of the waste arising from recovery operations intended to be disposed of in a landfill.
- Reporting by the waste authority concerning inspection activities carried out on solid waste service utilities, which are important for the regulator's activity, and, vice versa, namely the latter reporting situations which may indicate the presence of infringements.

The responsibilities of those two entities, the waste services regulator and the waste authority, may and should therefore be distinct but complementary, without the risk of overlaps or omissions.

15.6 ARTICULATION WITH THE PUBLIC HEALTH AUTHORITY

The public health authority normally has responsibility for the general coordination and planning of activities promoting health, preventing disease, and providing health care, with duties involving legislation, guidance, coordination and monitoring. It should intervene in the defence of public health, in the prevention of diseases and in health promotion and protection, as well as controlling risk factors in situations which may cause or increase serious harm to the health of the citizens or that of population clusters, as may be the case with the water and waste services.

The relationship of the water and waste services regulator with the health authority should include:

- Articulation in the implementation of public policies for water and waste services with public policies regarding public health.
- Articulation in the preparation of legislation and regulations for the water and waste services with public health legislation and regulations.
- Articulation in the collection, validation, processing and dissemination of water and waste services information with public health information.
- Articulation in boundary zones concerning drinking water quality regulations, in terms of ensuring the water quality is in conformity with legal requirements, which should be the responsibility of the regulator, and health surveillance, correlating this with issues of public health, for example with regard to epidemiological studies, which should be the responsibility of the health authority.
- Assessment of the risk regarding the protection of human health by the health authority in situations involving non-compliance with the quality parameters laid down in legislation for the quality of water for human consumption, which have been detected by the utility or by the regulator.
- Identification by the health authority of the cause/effect relationship between the water quality control results and water-borne diseases, thus allowing the regulator to either intensify or modify its intervention.

The responsibilities of those two entities, the water and waste services regulator and the public health authority, may and should therefore be distinct but complementary, without the risk of overlaps or omissions.

15.7 ARTICULATION WITH THE CONSUMER PROTECTION AUTHORITY

The consumer protection authority normally has the responsibility of contributing towards the drawing up, definition and carrying out of transversal consumer protection policy, with the aim of ensuring a high level of protection, particularly through monitoring the activity of consumer associations, arbitration centres for consumer conflicts, other extrajudicial resolution mechanisms for these conflicts and consumer information centres.

It should be noted while the responsibility of this authority in the promotion of consumer protection takes place within the framework of services in general, particularly essential public services, the economic regulation and quality of service under the responsibility of the regulator seeks to safeguard the rights and interests of all users of water and waste services, which corresponds to a wider group than that of just consumers (it includes all users, particularly commercial and industrial bodies). It is also carried out within a more integrated perspective, with concerns for the economic and also environmental sustainability of the services.

The relationship of the water and waste services regulator with the consumer protection authority should include:

- Articulation in the implementation of public policies for water and waste services with public policies regarding consumer protection.
- Articulation in the preparation of legislation and regulation for the water and waste services with consumer protection legislation and regulations.
- Articulation in the collection, validation, processing and dissemination of water and waste services information with consumer protection information, particularly concerning complaints supervised by the regulator.
- Articulation of boundary zones concerning the behavioural regulation of the utilities providing water and waste services to users, particularly legal and contractual regulation, economic regulation, quality of service regulation, drinking water quality regulation and user interface regulation.
- Joint promotion of institutionalised arbitration of consumer disputes.

The responsibilities of those two entities, the water and waste services regulator and the consumer protection authority, may and should therefore be distinct but complementary, without the risk of overlaps or omissions.

15.8 ARTICULATION WITH THE COMPETITION AUTHORITY

The water and waste services regulator should be subject to a principle of subsidiarity with regard to the competition authority. Its function is also to promote and defend competition, to respond to gaps in the market, besides ensuring the carrying out of public interest goals not necessarily ensured by the market, starting with ensuring the on-going and unbroken supply of certain goods and services essential to the community.

One of the tasks of the water and waste services regulator consists of promoting competition and contributing to the construction of a market from situations involving natural or legal monopoly, thus ensuring the progressive opening of the sectors and the development of healthy competition.

It should be noted that the progressive implementation of competition in the regulated sectors tends to concomitantly reduce the role of the water and waste services regulator in relation to the competition authority, with it being possible to conceptually state that sectors regulation has met its final goal when it is no longer necessary for society. However, in practical terms, the disappearance of the water and waste services regulator is not anticipated, even if theoretically possible within a long-term perspective.

The fact that the public water and waste services operate under a natural or legal monopoly and under the scope of exclusivity of rights does not exclude the need to consider measures to introduce competition.

The management of these services may allow for the existence of competition and access to the market, particularly when the service holders (State, municipalities, etc.) decide to involve private utilities to manage systems through a delegation or a concession contract, to participate in the capital of companies integrated in the State-owned business sector or the simple provision of services, typically infrastructure operation through a services provision contract.

In any of the described situations, competition should be ensured and maximised through public procurement rules, and it assumes special importance for issues such as:

- Non-discrimination, transparency and equal treatment within the scope of public procurement procedures, which seek to enable the participation of the greatest number of competitors in conditions involving equality of opportunities.

- The existence of limits to the length of contracts, which seek balance between a guarantee in terms of the recovery of investment, which is a factor in terms of the attractiveness of the contract for private bodies, and an increase in the instances of competition within the market.
- The existence of limits on the alteration of contracts, so as to prevent the distortion of the rules and underlying principles which led to the choice of the winning bidder.

In some service provision markets providing support to the management of water and waste systems competition problems may arise through the existence of significant buyer power on the part of the contracting entities, resulting from the existence of an oligopsony. This may strongly restrict these markets, for example those involving construction, the supply of products and equipment and public infrastructure consultants.

It should be noted that the utilities authorised by law to manage services of general economic interest or which have the form of a legal monopoly should remain submitted to competition rules, insofar as this does not constitute an obstacle to their compliance, in law or in fact, with the particular mission which has been entrusted to them. This issue is particularly important with regard to the application of the State aid system.

Finally, the unbundling of activities may be considered, similar to that which was carried out in the electricity and rail transportation sectors, to reduce the scope of the monopoly and liberalise certain activities.

The competition authority is normally responsible for the regulation of market competition, including the public drinking supply of water, waste water management, and solid waste management services and the various associated markets. The application of competition rules should be ensured, with regard to principle of the market economy and that of free competition, taking into account the efficient functioning of the markets, a high level of technical progress and the achievement of the greatest benefit to the users.

The relationship of the water and waste services regulator with the competition authority should include:

- Articulation in the implementation of public policies for water and waste services with public policies regarding competition.
- Articulation in the preparation of legislation and regulation for the water and waste services with competition legislation and regulations.
- Articulation in boundary zones concerning the legal and contractual regulation of the utilities throughout their life-cycle, specifically through analysing the tendering and contracting processes, contract modifications, contract terminations, and reconfigurations and mergers of systems.
- Articulation in the analysis of merger and reconfiguration operations of the utilities.
- Articulation in the detection, prevention and restriction of any concerted practices in public procurement procedures.
- Articulation in the identification, prevention and suppression of any abuses of a dominant position on the part of the water and waste services utilities within the framework of carrying out their secondary and supplementary activities concerning public services.
- Articulation in the detection, prevention and restriction of commercial practices potentially harmful for users.
- Articulation in assessing the applicability of any State aid system to the water and waste services.
- Reporting by the competition authority concerning inspection activities carried out on water and waste service utilities, which are important for the regulator's activity, and, vice versa, namely the latter reporting situations which may indicate the presence of competition infringements.

The responsibilities of those two entities, the water and waste services regulator and the competition authority, may and should therefore be distinct but complementary, without the risk of overlaps or omissions.

15.9 COOPERATION BETWEEN REGULATORY AUTHORITIES

Regulators have become an essential element of modern governance. For this reason, a growing number of regulators have been created in recent years.

According to the information available to ERSAR, and taking into account the difficulty in identifying what is or what actually is not a regulator, there currently exist more than one hundred and fifty (more precisely 167) water services regulators throughout the world, with different profiles, powers and levels of independence, geographically distributed throughout all the continents (Figure 15.1), where countries with regulatory authorities are in dark.

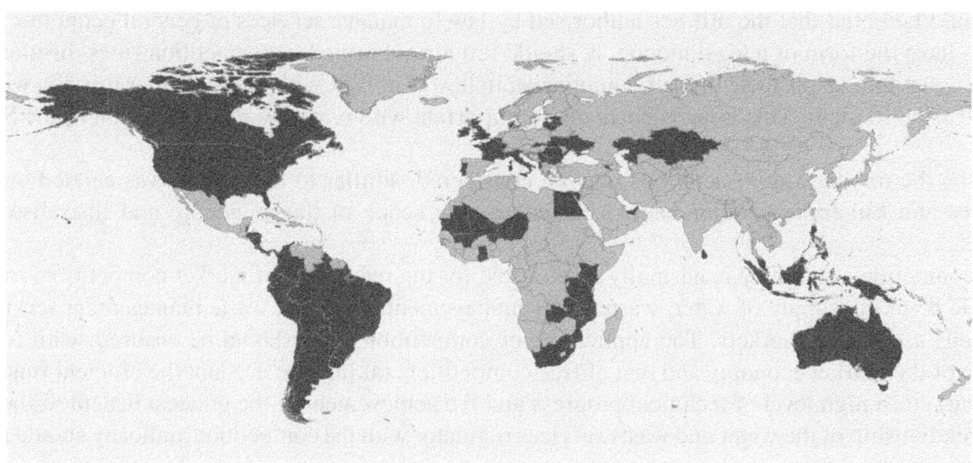

Figure 15.1 Water services regulators throughout the world.

In Europe 28 regulators were identified, mainly national regulatory authorities, covering Albania, Armenia, Belgium, Bulgaria, Croatia, Denmark, England and Wales, Estonia, France, Greece, Hungary, Ireland, Italy, Kazakhstan, Kosovo, Latvia, Lithuania, Malta, Northern Ireland, Portugal, Romania, Scotland, Slovakia and Ukraine, which are listed below:

- *Agency of the Republic of Kazakhstan on Regulation of Natural Monopolies (AREM) – Kazakhstan*
- *Commission for Energy Regulation (CER) – Ireland*
- *Commission for the Supervision of Water Resources (CVRI) – Italy*
- *Croatian Energy Regulatory Agency (CERA) – Croatia*
- *Danish Competition and Consumer Authority (KFST) – Denmark*
- *Danish Nature Agency (NST) – Denmark*
- *Estonian Competition Authority (ECA) – Estonia*
- *Flemish Environment Agency (VMM) – Belgium*
- *Hungarian Energy and Public Utility Regulatory Authority (HEA) – Hungary*
- *Malta Resources Authority (MRA) – Malta*
- *Ministère de l'Ecologie, du Développement Durable et de l'Energie, Direction de l'Eau et de la Biodiversité Bureau de l'Économie et de la Planification (DEB) – France*
- *National Commission for Energy Control and Prices National Control Commission for prices and energy (NCC) – Lithuania*

- *National Regulatory Authority for Municipal Services (ANRSC) – Romania*
- *Northern Ireland Authority for Utility Regulation (NIAUR) – Northern Ireland*
- *Office National de l'Eau et des Millieux Aquatiques (ONEMA) – France*
- *Public Services Regulatory Commission of the Republic of Armenia (PSRC) – Armenia*
- *Public Utilities Commission (SPRKPUC) – Latvia*
- *The Italian Regulatory Authority for Electricity Gas and Water (AEEGI) – Italy*
- *Regulatory Office for Network Industries (URSO) – Slovakia*
- *Special Secretariat for Water, Ministry of Environment, Energy and Climate Change – Greece*
- *State Committee for Water Management of Ukraine (SCWRM) – Ukraine*
- *State Energy and Water Regulatory Commission (SEWRC) – Bulgaria*
- *Water and Waste Regulatory Office (WWRO) – Kosovo*
- *Water and Waste Services Regulator (ERSAR) – Portugal*
- *Water and Waste Services Regulator (ERSARA) – Portuguese Autonomous Region of the Azores*
- *Water Industry Commission For Scotland (WICS) – Scotland*
- *Water Regulatory Authority (ERRU) – Albania*
- *Water Services Regulation Authority (OFWAT) – England and Wales*

In North America 45 regulators were identified, all of them regional, with 4 in Canada and 41 in the United States. Regional regulators are the model in both countries. They are listed below:

- *Alabama Public Service Commission (PSC) – USA – Alabama*
- *Alaska Regulatory Commission (RCA) – USA – Alaska*
- *Alberta Energy and Utilities Board (EUB) – Canada – Alberta*
- *Arizona Corporation Commission (AZCC) – USA – Arizona*
- *Arkansas Public Service Commission (APSC) – USA – Arkansas*
- *California Public Utilities Commission (CPUC) – USA – California*
- *Colorado Public Utilities Commission (PUC) – USA – Colorado*
- *Connecticut Department of Public Utility Control (DPUC) – USA – Connecticut*
- *Delaware Public Service Commission (PSC) – USA – Delaware*
- *Department of Public Utilities (DPU) – USA – Massachusetts*
- *Florida Public Service Commission (PSC) – USA – Florida*
- *Guam Public Utilities Commission (PUC) – USA – Guam*
- *Hawaii Public Utilities Commission (PUC) – USA – Hawaii*
- *Idaho Public Utilities Commission (PUC) – USA – Idaho*
- *Illinois Commerce Commission (ICC) – USA – Illinois*
- *Indiana Utility Regulatory Commission (IURC) – USA – Indiana*
- *Iowa Utilities Board (IUB) – USA – Iowa*
- *Kentucky Public Service Commission (PSC) – USA – Kentucky*
- *Louisiana Public Service Commission (LPSC) – USA – Louisiana*
- *Maine Public Utilities Commission (MPUC) – USA – Maine*
- *Maryland Public Service Commission (PSC) – USA – Maryland*
- *Mississippi Public Service Commission (MSPSC) – USA – Mississippi*
- *Missouri Public Service Commission (PSC) – USA – Missouri*
- *Montana Public Service Commission (PSC) – USA – Montana*
- *Nevada Public Utilities Commission (PUCN) – USA – Nevada*
- *New Hampshire Public Utilities Commission (NHPUC) – USA – New Hampshire*
- *New Jersey Board of Public Utilities (NJBPU) – USA – New Jersey*

- *New York State Public Service Commission (PSC) – USA – New York State*
- *North Carolina Utilities Commission (NCUC) – USA – North Carolina*
- *Nova Scotia Utility and Review Board (NSUARB) – Canada – New Scotia*
- *Oregon Public Utility Commission (PUC) – USA – Oregon*
- *Pennsylvania Public Utility Commission (PUC) – USA – Pennsylvania*
- *Prince Edward Island Regulatory and Appeals Commission (IRAC) – Canada – Prince Edward Island*
- *Public Service Commission of Wisconsin (PSC) – USA – Wisconsin*
- *Public Utilities Commission of Ohio (PUCO) – USA – Ohio*
- *Rhode Island Public Utilities Commission (RIPUC) – USA – Rhode Island*
- *South Carolina Office of Regulatory Staff (ORS) – USA – South Carolina*
- *Tennessee Regulatory Authority (TRA) – USA – Tennessee*
- *The Public Utilities Board (PUB) – Canada – Manitoba*
- *Utah Public Service Commission C) – USA – Utah*
- *Vermont Public Service Board (PSB) – USA – Vermont*
- *Virgin Islands Public Services Commission (PSC) – USA – Virgin Island*
- *Virginia State Corporation Commission (SCC) – USA – Virginia State*
- *West Virginia Public Service Commission (PSC) – USA – West Virginia*
- *Wyoming Public Service Commission (PSC) – USA – Wyoming*

In South and Central America 62 regulators were identified, covering Anguilla, Argentina, Bahamas, Barbados, Belize, Bolivia, Brazil, Chile, Colombia, Costa Rica, Ecuador, Guyana, Honduras, Jamaica, Nicaragua, Panama, Paraguay, Peru, Trinidad and Tobago and also Uruguay, which are identified below. Regional regulators dominate in Brazil, with 30, and in Argentina, with 13:

- *Authority for the Supervision and Control of Water and Basic Sanitation (AAPS) – Bolivia*
- *Fair Trading Commission (FTC) – Barbados*
- *Goiana Agency for the Regulation, Control and Inspection of Public Services (AGR) – Brazil*
- *Intermunicipal Agency for the Regulation, Control and Inspection of Municipal Public Services of Médio Vale do Itajaí (AGIR) – Brazil*
- *Intermunicipal Regulatory Agency for Waste water (ARIS) – Brazil*
- *Municipal Agency for the Regulation of Public Services of Teresina (ARSETE) – Brazil*
- *Municipal Agency for the Regulation of the Sanitation Services of Cachoeiro de Itapemirim (AGERSA) – Brazil*
- *Municipal Agency for the Regulation of Water and Sewage Services of Joinville (AMAE) – Brazil*
- *Municipal Water and Sanitary Sewage Agency of Cuiabá (AMAES) – Brazil*
- *National Superintendence of Sanitation Services (SUNASS) – Peru*
- *Nicaraguan Institute for Aqueducts and Sanitary Sewage (INAA) – Nicaragua*
- *Organisation for Water Control of Buenos Aires (OCABA) – Argentina*
- *Provincial Body for Water and Sanitation of the Province of Mendoza (EPAS) – Argentina*
- *Provincial Department of the Waters of Río Negro (DPA) – Argentina*
- *Public Utilities Commission (PUC) – Bahamas*
- *Public Utilities Commission (PUC) – Belize*
- *Public Utilities Commission (PUC) – Guyana*
- *Public Utilities Commission of Anguilla (PUC) – Anguilla*
- *Regulated Industries Commission (RIC) – Trinidad and Tobago*
- *Regulation Committee for Drinking Water and Basic Sanitation (CRA) – Colombia*

- *Regulatory Agency for Basic Sanitation and Road Infrastructure of Espírito Santo (ARSI) – Brazil*
- *Regulatory Agency for Basic Sanitation Services of the Municipality of Natal (ARSBAN) – Brazil*
- *Regulatory Agency for Basic Sanitation Services of the State of Santa Catarina (AGESAN) – Brazil*
- *Regulatory Agency for Energy and Basic Sanitation of the State of Rio de Janeiro (AGENERSA) – Brazil*
- *Regulatory Agency for Outsourced Public Services for the State of Pernambuco (ARPE) – Brazil*
- *Regulatory Agency for Public Services of Santa Catarina (AGESC) – Brazil*
- *Regulatory Agency for Public Services of the State of Acre (AGEAC) – Brazil*
- *Regulatory Agency for Public Services of the State of Alagoas (ARSAL) – Brazil*
- *Regulatory Agency for Public Services of the State of Ceará (ARCE) – Brazil*
- *Regulatory Agency for the Public Services Granted by the State of Amazonas (ARSAM) – Brazil*
- *Regulatory Agency for the State of Paraíba (ARPB) – Brazil*
- *Regulatory Agency for the Waste water Services of the Catchment Areas of the Piracicaba, Capivari and Jundiaí Rivers (ARES–PCJ) – Brazil*
- *Regulatory Agency for the Water and Sewage Services of Mauá (ARSAE) – Brazil*
- *Regulatory Agency for the Waters of Tubarão (AGR) – Brazil*
- *Regulatory Agency for Waste water and Energy of the State of São Paulo (ARSESP) – Brazil*
- *Regulatory Agency for Water Supply Services and Sanitation Systems of the State of Minas Gerais (ARSAE) – Brazil*
- *Regulatory Agency for Waters, Energy and Basic Sanitation of the Federal District (ADASA) – Brazil*
- *Regulatory Agency of Guaratinguetá (ARSAEG) – Brazil*
- *Regulatory and Control Body for the Drinking Water and Sanitation Concession of Guayaquil (EMAPAG) – Ecuador*
- *Regulatory Authority for Public Services (ARESEP) – Costa Rica*
- *Regulatory Body for Drinking Water and Sanitation Services (ERSAPS) – Honduras*
- *Regulatory Body for Public Services (ERSP) – Panama*
- *Regulatory Body for Sanitary Services (ERSSAN) – Paraguay*
- *Regulatory Body for Sanitary Services of Santa Fe (ENRESS) – Argentina*
- *Regulatory Body for the Public Services of Córdoba (ERSeP) – Argentina*
- *Regulatory Body for the Public Services of the Province of Salta (EnReSP) – Argentina*
- *Regulatory Body for the Public Works and Services of the Province of Formosa (EROSP) – Argentina*
- *Regulatory Body for the Water and Sewage Service of Tucumán (ERSACT) – Argentina*
- *Regulatory Body for Water and Sanitation of Buenos Aires (ERAS) – Argentina*
- *Regulatory Body for Water and Sewage Services of Santiago del Estero (ERSAC) – Argentina*
- *Single Controlling Body for the Privatisations of the Province of La Rioja (EUCOP) – Argentina*
- *Single Regulatory Body for the Public Services of Buenos Aires – Argentina*
- *State Regulatory Agency for Basic Sanitation of Bahia (AGERSA–BA) – Brazil*
- *State Regulatory Agency for Outsourced Public Services of Rio Grande do Sul (AGERGS) – Brazil*
- *State Regulatory Agency for Outsourced Public Services of the State of Mato Grosso (AGER) – Brazil*
- *State Regulatory Agency for Public Services of Mato Grosso do Sul (AGEPAN) – Brazil*
- *State Regulatory and Control Agency for the Public Services of Pará (ARCON) – Brazil*
- *Superintendence of Public Services and Other Concessions of the Province of Jujuy (SUSEPU) – Argentina*
- *Superintendence of Sanitary Services (SISS) – Chile*
- *The Office of Utilities Regulation (OUR) – Jamaica*

- *The Regulatory Unit for Energy and Water Services (URSEA) – Uruguay*
- *Tocantinense Agency for the Regulation, Control and Inspection of Public Services (ATR) – Brazil*

There were 13 regulators identified in Africa, mainly national ones, covering South Africa, Gambia, Ghana, Cape Verde, Mauritania, Mozambique, Niger, Kenya, Rwanda, Tanzania, Zambia, Mali and Malawi, which are listed below:

- *Autorite de Regulation Multisectorielle (ARM) – Niger*
- *Autoritee Regulation (ARE) – Mauritania*
- *Economic Regulation Agency (ARE) – Cape Verde*
- *Energy and Water Regulatory Authority (EWURA) – Tanzania*
- *Gambia Public Utilities Regulatory Authority (PURA) – The Gambia*
- *Malawi Water and Energy Regulatory Authority (MWERA) – Malawi*
- *Mali Electricity and Water Regulatory (CREE) – Mali*
- *National Water Supply and Sanitation Council (NWASCO) – Zambia*
- *Public Utilities Regulatory Commission (PURC) – Ghana*
- *Rwanda Utility Regulatory Agency (RURA) – Rwanda*
- *South Africa Water Research Commission (WRC) – South Africa*
- *Water Services Regulatory Board (WASREB) – Kenya*
- *Water Supply Regulatory Board (CRA) – Mozambique*

There were 7, primarily national, regulators identified in Asia, namely Egypt, United Arab Emirates, Philippines, Indonesia, the Maldives, Sri Lanka and Malaysia, which are listed below:

- *Abu Dhabi Regulation and Supervision Bureau (RSB) – United Arab Emirates*
- *Environmental Protection Agency (EPA) – the Maldives*
- *Jakarta Water Supply Regulatory Body (JWSRB) – Indonesia*
- *National Water Services Commission (SPAN) – Malaysia*
- *Public Assessment of Water Services (PAWS) – Philippines*
- *Public Utilities Commission of Sri–Lanka (PUCSL) – Sri Lanka*
- *Water and Waste water Regulatory Agency (EWRA) – Egypt*

There were 12 regulators identified in Oceania, mainly in Australia, with 11, but also in Papua New Guinea, which are listed below:

- *Australian Competition and Consumer Commission (ACCC) – Australia*
- *Economic Regulation Authority of Western Australia (ERA) – Australia*
- *Essential Services Commission of South Australia (ESCOSA) – Australia*
- *Government Prices Oversight Commission of Tasmania (GPOC) – Australia*
- *Independent Competition and Regulatory Commission (ICRC) – Australia*
- *Independent Competition and Regulatory Commission of Australian Capital Territory (ACT) – Australia*
- *Independent Competition and Regulatory Commission of New South Wales (IPART) – Australia*
- *Northern Territory Utilities Commission – Australia*
- *Papua New Guinea Water Board – Papua New Guinea*
- *Queensland Competition Authority (QCA) – Australia*
- *Queensland Water Commission (QWC) – Australia*
- *Victorian Essential Services Commission (ESC) – Australia*

With regard to solid waste management services, there currently appear to be very few explicit regulators throughout the world, with known regulators existing in Brazil, Hungary, Portugal and in Kosovo.

These regulatory bodies have similar goals and deal with problems which are often common. They may thus benefit from processes of cooperation, as shown for example in bilateral protocols, promoting specific actions for the exchange of experiences, exchange of information and documentation and training and technical capacity building activities. To implement these protocols, the parties may for example provide technical staff to monitor, develop and participate in joint activities drawn up by the parties and mutually make information available, which are necessary elements and data for the development of established activities as well as welcome collaborators from the other party in their facilities.

Supplementary to this bilateral relationship, and with the goal of creating common discussion and reflection forums for the different regulatory bodies, regional networks have been set up in various parts of the world, such as:

- *African Forum for Utility Regulators (AFUR).*
- *Association of Regulatory Bodies for Drinking Water and Sanitation for the Americas (ADERASA).*
- *East Asia and Pacific Infrastructure Regulatory Forum (EAPIRF).*
- *Eastern and Southern Africa Water and Sanitation Regulators Association (ESAWAS).*
- *Energy Regulators Regional Association (ERRA), network of energy and water regulatory bodies primarily from the Central European and Eurasian region.*
- *European Water Regulators Network (WAREG).*

Also OECD created a Network of Economic Regulators (NER) and promotes studies to strength regulatory frameworks and governance in the water sector.

In the countries in which regulation takes on a regional structure, for example by states, there are also national networks, such as:

- *Brazilian Association of Regulation Agencies (ABAR).*
- *Canada's Energy and Utility Regulators (CAMPUT).*
- *National Association of Regulatory Utility Commissioners (NARUC), in the United States of America.*

There are also specific regulatory networks for water quality, such as:

- *European Network of Drinking Water Regulators (ENDWARE).*
- *International Network of Drinking–Water Regulators (REGNET).*

In situations involving the construction of single markets, for example the European one, there is a reinforced need for effective coordination of similar national regulatory bodies, and it is even conceptually possible to consider partial or total ways of integrating their regulatory function within such economic spaces.

In analysing the geographical distribution of regulation throughout the world's continents, it became evident that regulation of water services (albeit less evident in waste services) is increasingly taking on universal importance, particularly due to the need, generalised to many countries, to implement more effective public policies for the provision of these essential public services.

The regulatory bodies are increasingly seen as important agents in the implementation of these public policies, and thus managing better the relationship between governments, utilities and users, bringing greater rationality and equity amongst all sectors' stakeholders.

A greater level of cooperation is therefore required, promoting a network of regulatory bodies at the world level, to enlarge the discussion of models, strategies and regulatory procedures, as well as their trends, which can have a great impact on the provision of public water services. This is the goal for holding the *1st International Water Regulators' Forum,* organised by the *International Water Association in partnership with ERSAR,* as part of the *IWA World Water Congress & Exhibition* held in Lisbon, Portugal, in September 2014.

This network, currently of an informal nature, has brought together a large number of bodies which carry out regulatory activities for drinking water supply and waste water management services, sometimes together with other services, such as energy, telecommunications, transportation or solid waste management, and which seeks to contribute towards the international harmonisation and dissemination of best regulatory practices. It forms one exceptional opportunity to discuss regulatory principles, regulatory independence and public accountability, as well as legal and contractual and economic regulation, regulation of quality of service, quality of water for human consumption and interaction with users. It is also an opportunity to reinforce articulation with the environmental, water resources, waste management, public health, consumer protection and competition authorities.

In Portugal ERSAR, along with the Centre for Urban and Regional Systems (CESUR) of the Instituto Superior Técnico in 2011 jointly published the study 'The regulation of the water supply and waste water services – An international perspective', written by Rui Cunha Marques, with contributions from Pedro Simões, João Simão Pires, João Almeida and Tiago Neves. Its aim was to bring together experiences from the five continents regarding the provision of public water services and the regulatory models adopted. Information was collected on around 60 countries, thus enabling the comparison and identification of interesting case studies and potential good practices. This study brings together, in a single publication, systematic and comparable information on the legal and institutional framework of each country, the structure of the markets, the identification of the main stakeholders and numbers for the sectors, including tariff structures, the regulatory models adopted, and the various mechanisms used in the assessment of the quality of service and compliance with the obligations of public service.

15.10 SUMMARY

The present chapter considered the desirable articulation of the regulator with the main stakeholders in the public drinking water supply, waste water management and solid waste management services.

It is considered that conditions do not exist for effective and efficient performance by the regulators without the suitable articulation with the main stakeholders. That articulation must be focused on achieving synergies, transparency of procedures, and additional collection of information, which are success factors for regulation, giving also greater legitimisation to the regulator.

The following chapter will present the main conclusions of this book.

Chapter 16

Conclusions

16.1 INTRODUCTORY NOTE

This chapter summarises, within the integrated regulatory approach (model RITA-ERSAR) that is being advocated, the main features of the drinking water supply, waste water management and solid waste management public services and identifies what is considered to be the responsibilities of political power, regulatory authorities, utilities, users and society as a whole. This ensures that these services can be provided with universal access, continuity and quality of service and efficiency and price equity, thus constituting an important factor for development and social balance.

16.2 WATER AND WASTE SERVICES

The public water supply, waste water and solid waste management to populations constitute public services, of a structural character. They are essential to the general welfare, public health and collective security of the population, economic activities and the protection the environment. Consequently they have a fundamental importance in societies and are usually classified as services of general economic interest.

Therefore:

- They should obey a set of public service obligations, which are universal access to the services, the suitability of the services in terms of quantity, quality and continuity, the structural and operational efficiency, the suitability and fairness of the prices and the adoption of rules of good practice.
- They have important and sometimes specific characteristics such as the fact that they are irreplaceable services that deal with heterogeneous products, which have potential economies of scale, of scope and of process, and tend to be regional, and use assets designed for peak situations, which are high cost, of long duration and high immobilisation and show a long period for the recovery of invested capital and low elasticity between price and demand.
- They are generally provided under a natural or legal monopoly, and there is therefore no clear incentive for the search for greater efficiency and effectiveness by the utilities with an increasing prevalence of risks to the users.

- These sectors contains numerous and diverse types of stakeholders, which can be divided into several groups, such as the public administration, service holders, the utilities, entities providing services and civil society.
- A public water system can be considered as usually having the infrastructure components of abstraction, treatment, pumping, supply, storage, distribution and the recovery of by-products.
- A waste water system can be considered as usually having the infrastructure components of abstraction, pumping, transportation, treatment, the discharge of treated waste water, the treatment of sludge and the reuse and recovery of by-products. A storm water system can normally be considered to have as infrastructure components drainage, retention, treatment and the discharge of storm water.
- A solid waste management system can normally be considered to have as infrastructure the collection, transportation, storage, sorting, recovery and disposal of waste.
- The services of water supply for human consumption and waste water, components of the urban water cycle, are strongly dependent on water resources, as these constitute the raw material base for the production of drinking water and are also in most situations the final destination for treated, or untreated, waste water. Indeed, suitably treated waste water is in general disposed of in the water resource and, after extensive dilution and natural purification, reused indirectly through the water cycle, preferably with appropriate treatment, or used directly for various purposes such as agriculture, as a water resource and source of nutrients.
- The services of solid waste management are also heavily dependent on the environment because it forms the raw material base for the production of consumer goods which give rise to waste and which is also, in most situations, the final destination of solid waste, treated or not. Indeed, waste is disposed of in the environment and subsequently reused indirectly in the long term through the natural life cycle of materials or, more directly and in the short term, reused after treatment or by recycling for various purposes, as a raw material or resource.
- The most important challenges for the water and waste services tend to exist primarily at various levels of the management of the public services, in the interface of the service with its users, considering public health and the environment, and also in relation to the models of governance of the services.

16.3 RESPONSIBILITIES OF THE POLITICAL POWER

It is recommended, given these considerations, that political power, as a key stakeholder in these sectors, ensures the implementation of appropriate public policies for the provision of the drinking water supply, waste water management and solid waste management services provided to the population, in particular within the framework of the Millennium Development Goals, which set targets for the water services in terms of population coverage, and the recent United Nations resolution, which declared water and sanitation services to be human rights.

These policies should be based on an integrated approach involving various aspects:

- Adoption of strategic plans for the sectors, at a national or regional level and in the medium term, which embodies the vision of governments for the sectors and society.
- Definition of the legislative framework that considers the legal scheme for the services and regulation and also legislation on water, waste, environment, consumer protection and competition.
- Definition of the appropriate institutional framework, with clear allocation of responsibilities amongst the various public entities involved, essential for good sectors performance.
- Definition of the governance of services, defining the models that can be used, for example direct management, delegated management and concessioned management, depending from the political willingness and decision in each country and context.

- Definition of realistic access goals, namely the population to have these services available, as well as the objectives of quality of service.
- Definition of a tariff policy for services that promotes a gradual recovery trend for costs, compatible with the economic capacity of the population and which protects the most economically disadvantaged.
- Definition of a tax policy for the services which creates suitable economic incentives in terms of behaviour, management and use of water and waste services.
- Availability and efficient management of important financial resources, from own funds or from the cooperation and development support funds.
- Construction of the infrastructure necessary for the provision of services, using appropriate technology, naturally involving major investments.
- Improvement in the structural efficiency of the services, with an optimised territorial organisation, and also the efficiency of operation by the utilities.
- Human resource capacity building in number and technical skills to ensure the general quality of the service.
- Promoting research and development in areas related to the services, creating endogenous knowledge and thus ensuring increased national technical autonomy.
- Development of the business sector, with the strengthening of its capacity in national and possibly international markets, by creating jobs and wealth.
- Introduction of competition into the services to encourage innovation and technical progress and, therefore, increasing the efficiency and quality of its provision.
- Awareness and participation of users regarding the services, so that there is greater and more fruitful civic participation.
- Providing reliable information on services, both to support the definition of public policies and business strategies and to ensure greater transparency in the provision of water and waste services provided to society.

The regulation should be seen as a component of public policies on water and waste. Although this is only one component among many, it does however play a very important role, in so far as it is the promoter or even responsible for the control of most of the other components.

It is considered to be of general interest that governments, at central, regional and local levels, assume a clear focus on strengthening the regulation of essential public services, which are one of the mainstays of citizenship, particularly because they function in natural or legal monopoly markets, where there is no natural incentive to seek out greater efficiency and effectiveness by the utilities and where the prevalence of these risks to the users are higher.

As such, the regulation framework should be reviewed in order to:

- Expand the solution of having dedicated regulatory authorities in the sectors to all situations in which these assumptions have been verified, gradually substituting in these cases the common instances of the direct administration of the State.
- Ensure the adequate level of organic, functional and financial independence of the regulatory authorities, to ensure the stability and autonomy of their bodies and freedom of decision of these in their legally defined activity area, except for judicial review.
- Establish the necessary mechanisms to ensure accountability and public scrutiny of regulatory bodies, particularly with regard to transparency of their action, reinforced procedural requirements for preparing and publishing periodical reports of their activities and, where appropriate, for presentation to the relevant parliamentary committees.

- Define in a framework law a minimum common reference to the basic principles that should guide the organisation, powers and functioning of regulators, thereby providing robustness to the system, without prejudice to there being a wide margin of discretion in the status of each regulatory authority, due to the specifics of each case.

Regulators have been forming an essential element of modern governance, providing greater separation between the areas of politics and economics, as a key instrument of the very constitution of the competitive market in sectors previously outside its logic, fostering a new regulatory culture less dependent on governments and more rational and more objective. With these regulators the State and the economy potentially have better governance.

The contribution of independent regulation to the modernisation of public administration and management and the economy can be strengthened, providing greater system-wide coherence, by harmonising solutions, eliminating bottlenecks and increasing the confidence of economic agents and users in the objectives and practice of regulation.

16.4 RESPONSIBILITIES OF REGULATORY AUTHORITIES

It is recommended that the regulatory authorities, materialized on one or more organizations, as major stakeholders in these sectors, ensure the implementation of regulatory models which form an integrated regulatory approach, for example for the public water supply, waste water management and solid waste management services, regulating both the sectors as a whole and the utilities individually, looking to find the optimal global solution for the populations.

An integrated regulatory approach for these water and waste services has been presented in a practical manner, nominated as the RITA-ERSAR regulation model, in which:

- The model should include a level of structural regulation for the sectors, in which the regulatory authority should contribute to the organisation, legislation, information and capacity building of the sectors, and a level of behavioural regulation for the utilities, in which the regulatory authority should carry out legal and contractual regulation, economic regulation, regulation of the quality of service, regulation of the quality of drinking water and regulation of the interface with users.
- The structural regulation of the sectors should contribute to the organisation of the sectors, particularly in terms of an optimal territorial organisation taking advantage of economies of scale, scope and process.
- It should also contribute to the legislative instruments for the sectors, with the goal of helping to clarify its operating rules, an aspect which is of course essential for the proper delivery of these services through instruments with a variable degree of external effectiveness, specifically laws, regulation and recommendations.
- It should also contribute to information for the sectors, developing and regularly disseminating accurate and accessible information to all stakeholders. It should be the responsibility of the regulatory authority to create a national information system and to consolidate a culture of concise, credible information which can be easily interpreted by all, extendable to all utilities, regardless of the forms of management adopted for the provision of services.
- Finally it should contribute to capacity building in the sectors, promoting research and development, creating innovation and endogenous knowledge and human resources capacity building with suitable technical and professional training to better carry out their activity, thus ensuring increased autonomy for the country's water and waste services.

- The behavioural regulation of the utilities should play a role in the legal and contractual area, with the aim of ensuring that all of the stages of its life-cycle, from the design stages, any tendering process, contracting, service management, contract amendment and termination, are carried out in strict compliance with legislation and any existing contract, as is the case in situations involving delegation and concessions.
- It should also play a role in the economic area, with the aim of ensuring the application of tariffs and a suitable tariff scheme, with appropriate behaviour and economic and financial efficiency from the utilities, promoting the rationalisation of prices to users at the same time as the economic and financial sustainability of the utility.
- It should also play a role in the area of quality of service, with the aim of ensuring the provision of a suitable quality of service to users by utilities under the terms of its goals and the applicable law.
- It should also play a role in the area of the quality of drinking water, with the goal of ensuring in a continuous manner the supply of an appropriate level of water quality by utilities, in accordance with applicable law, for the benefit of public health.
- It should also play a role in the area of the interface with users, in order to better ensure the protection of users' rights through compliance with legislation for consumer protection, defence of the right to make a complaint and subsequently improving the quality of the relationship of the utilities with the users.
- Regulatory authorities should act based on the principles of competence, exemption, impartiality, accountability and transparency.

16.5 RESPONSIBILITIES OF THE UTILITIES

It is recommended that the utilities, public and private, as key stakeholders in these sectors effectively and efficiently ensure suitable public water, waste water and solid waste management services to local populations, thus contributing to the development of society.

To carry this out the utilities, regardless of their management model, whether direct, by delegation, by concession, or other, should:

- Have a legal or contractual basis to formalise their responsibilities in providing their service.
- Have use of human, technical and financial resources and a suitable physical infrastructure in order to meet the population's needs in terms of services.
- Develop its activities within the framework of strategic plans for the sectors for the drinking water supply, waste water management and solid waste management and the respective legislation.
- Contribute to improving the structural efficiency of the sectors by making use of economies of scale through the physical interconnection of systems on a technically and economically appropriate scale, in which the benefits in terms of reducing unit costs and tariff harmonization are evident.
- Further promote the utilisation of economies of scope, where applicable, through the aggregation of the water supply systems with waste water systems, provided that the benefits are clear in terms of reducing unit costs.
- Also promote the use of economies of process, through the possibility, when applicable, of integration of the bulk and retail systems, provided the solutions found are technically rational and the benefits become apparent in terms of reduced unit costs.
- Encourage improvements in the operating efficiency of the utilities, which should seek to adopt, given the existing legislation, the most advisable-type of organisation, particularly in relation to staff, the functional content of information circuits, administrative routines, financial resources, planning, budget, control and quality assurance.

- Reach goals in terms of the population accessing the public water and waste services in a continuous manner with suitable quality of service, promoting the construction of infrastructure elements which are still necessary, with recourse to suitable solutions and technologies.
- Develop activities that contribute to green growth and in particular lead to waste minimization and recovery of by-products that result from its activity, including energy recovery of waste, waste water and sludge.
- Act in strict compliance with the legislative and contractual framework, particularly the legal framework for the sectors and environmental legislation, tariff structure, quality of service and quality of water, consumer protection and competition.
- Ensure collaborative models of governance with other entities, leveraging the synergies of shared management and the convergence of objectives, with mutual advantages and for the benefit of users.
- Act in cooperation with due regard for the powers of the public entities involved in the sectors, in particular the regulatory authority and the environmental authorities, water resources, health, consumer protection and competition.
- Implement a pricing policy under the principle of user-payer to promote a recovery of costs in a very efficient environment and compatible with the economic capacity of the population, particularly the most disadvantaged.
- Ensure optimal use of any funding available, prioritising projects that maximise the benefit of investments, significantly improving the quality of service and environmental performance and enabling acceptable tariffs according to the level of economic and social development of the population.
- Contribute to the capacity building of human resources and sectors innovation in cooperation with other bodies, essential factors for overall quality assurance, including creating endogenous knowledge and thus ensuring increasing national autonomy.
- Promote the development of the national, regional and local business sector, particularly through partnerships and the contracting of service providers, thus creating better conditions for the development of national knowledge and, hence, strengthening the capacity building of the national and possibly international business sector by creating jobs and wealth.
- Contribute to increasing the introduction of competition in the sectors, motivating innovation and technical progress and thus increasing the efficiency and the quality of the services while minimising their monopolistic characteristics and the risk of abuse of a dominant position and other anti-competitive practices contrary to the interests of users.
- Ensure a proper user interface, the true recipients of these essential public services, part of them considered to be human rights.
- Keep track of information and conduct suitable and auditable accounting, in accordance with the requirements of the regulatory authority in particular, and provide reliable information both to support the specification of public policies and business strategies as well as to evaluate the service that is actually provided to society.

16.6 RESPONSIBILITIES OF THE USERS

It is also recommended that users, as major stakeholders in these sectors and final recipients of the public water supply, waste water and solid waste management services to local populations:

- Effectively exercise their rights, particularly regarding physical and economic access to services, their quality, the quality of drinking water, information on the services, the possibility of making a complaint about the services and participating at adequate levels in decisions, and assuming their corresponding duties.

- Are citizens and organizations aware of and concerned with the water and waste services, so that their behaviour can minimise any adverse impact on public health and the environment, such as avoiding the potential contamination of water sources and the reduction in quality and/or reliability of the water being distributed.
- Properly maintain home-based systems for water supply, waste water and collection of waste, through appropriate procedures for the use and adoption of appropriate materials and equipment.

16.7 COMMON RESPONSIBILITIES

Finally it is recommended that all the stakeholders in these sectors, such as governments, public administrations, service holders and utilities, the entities providing services and civil society, particularly users, have an ongoing and open dialogue and share information on carrying out and maintaining suitable public water supply, waste water and solid waste management services to local populations.

16.8 FINAL NOTE

This chapter has presented the main conclusions, including a set of general recommendations addressed to the political powers, regulatory authorities, utilities, users and society in general with regard to public drinking water supply, waste water management and solid waste management services to local populations.

It is the author's profound conviction that if the political powers implement a suitable public policy that includes an integrated regulatory approach, if the utilities and regulatory authorities perform their activity appropriately and if users play their part in proactive citizenship, then the essential conditions have been met to promote the provision of public water and waste services with universal access, continuity and quality and efficiency and price equity, thus constituting an important factor for social and economic development.

Annex A

The evolution of water and waste services in Portugal in the last two decades

1. INTRODUCTORY NOTE

Since the definition in 1993 of a new public policy for the water and waste services, there has occurred in Portugal a very positive evolution in the provision of these essential public services, with an enormous increase in social well-being and its impact on public health and the environment, without prejudice to some aspects still in need of improvement.

Below is described the most important components of this new political policy which has been implemented in the last two decades in Portugal, as well as the results of this implementation, particularly in terms of the evolution of the public services of drinking water supply, waste water management and solid waste management, and the impact on environmental quality and on public health and also the impact of compliance with human rights in access to water and sanitation.

Regulation has played an important role in this process, insofar as it has promoted, enforced and monitored most of the remaining components of this public policy. It also contributes towards improving the transparency of these services, bringing more responsibilities and scrutiny and creating a more demanding society, and positive pressure for improving the performance of the sectors.

2. THE NEW PUBLIC POLICY FOR WATER AND WASTE SERVICES

2.1. Overview

In Portugal, as in any other country, the water and waste services are very complex and involve many areas, which are distinct but must be interconnected. It is necessary to bring together in a suitable way institutional, governance, management, planning, technical, economic, legal and environmental, public health, social and ethical instruments to ensure the suitable provision of these services.

In 1993, when in Portugal it was decided to define a new public policy which would ensure suitable water and waste services to the population, a global integrated approach was adopted, that is to say, a holistic one, involving various components, which are described below:

- Approval of strategic plans for the sectors;
- Definition of the legislative framework;
- Definition of the institutional framework;

- Definition of the governance of the services;
- Definition of the access targets and the quality of service goals;
- Definition of the tariff and tax policy;
- Provision and management of financial resources;
- Construction of the infrastructure;
- Improving structural and operational efficiency;
- Human resources capacity building;
- Promotion of research and development;
- Development of the business sector;
- Introduction of competition;
- Protection, awareness and involvement of users;
- Making information available.

It was thus believed that the successful implementation of a new public policy for water and waste services was dependent on the ability to manage the implementation of all these components at relatively the same time, ensuring an effective global and integrated approach. Trying to resolve the problem with just one or several of these components would lead to failure and certainly not allow the goals to be achieved in a sustained manner. A considerable effort would have been made without an overall perspective, which would have consumed resources without the expected results.

There follows an analysis in greater detail of each one of these components, based on the experience acquired in the implementation of this new public policy.

This policy was naturally based on a careful analysis of the evolution of the sectors in the past, learning from successes but mainly from the errors which had been committed.

In Portugal a large-scale structural study was carried out by the National Laboratory for Civil Engineering in 1993 entitled 'Support tools for a sustainable development policy regarding basic sanitation', which gave rise to an editorial series entitled 'Management of basic sanitation systems' which ran to 16 Volumes and was published in 1995. This study was carried out at the request of the then Director-General of the Environment, Artur Ascenso Pires, and provided one additional support for public policies that were defined for the water and waste services.

Recently, in 2011 ERSAR published the study 'History of public policies for water supply and sanitation in Portugal', written by João Howell Pato, which carried out a reconstitution and historic analysis of water supply and sanitation in Portugal from the end of the 19th century until the first decade of this century. It was also made a critical analysis of the facts, to support reflection concerning improvements to be introduced in the design, definition, implementation and assessment of policies for the sectors.

2.2. Adoption of strategic plans for the sectors

The Portuguese situation concerning water and waste services before 1993 was inacceptable regarding the expectations of the population and the country's development ambitions, with no date having been specified for a clear strategy to resolve these issues.

The political, social and economic importance attributed to these sectors in 1993 led to a political commitment to reorganise the sectors, involving appropriate strategies, expressed within a strategic plan for the sectors for drinking water supply and waste water management, and another strategic plan for the sector for the management of solid waste, both nationwide in scope and for the medium term, corresponding to the government's vision for these sectors and for society.

The strategic plan for drinking water supply and waste water management involved a first-generation (1993–1999), a second-generation (2000–2006), a third-generation (2007–2013) and now the start of a fourth generation (2014–2020). In a similar manner, the strategic plan for solid waste management had a first-generation (1993–1996), a second-generation (1997–2006), a third-generation (2007–2013) and now the start of a fourth generation (2014–2020).

These strategic plans involve stages characterising and diagnosing the current situation, defining intended goals, evaluating the corresponding investment needs, identifying the measures needed, implementing strategies and specifying the monitoring instruments.

One of the success factors was without doubt the stability over time of this public policy during these twenty years, nevertheless the natural rotation of parties in government, with their different policy options. Although subject to natural minor adaptations in terms of its path and evolution, the public policy for the water and waste services remained quite stable and was not subject to a process of having to take steps forward then steps backwards.

Like other stakeholders, the regulator has contributed to the formulation of this public policy, towards its rationalisation and the resolution of any malfunctions regarding the regulated services and towards the organisation of the sectors, promoting for example an increase in the efficiency and effectiveness of the water and waste services and the search for economies of scale, scope and process. It has also monitored the strategies adopted for the sectors, by accompanying their implementation and regularly reporting on their evolution and their constraints.

2.3. Definition of the legislative framework

In so far as public policy needs to be reflected in legislation, Portugal set up a new and modern legislative framework to cover the legal regime for water and waste services and their regulation, as well as tariff legislation, quality of service, quality of water and also the technical aspects. There was essentially a first generation of legislation in 1993 and a second generation in 2009. Additionally, the country improved modern legislation on water resource management, waste management, environmental management, consumer protection and competition.

The sectors' stakeholders, particularly the government, public administration, State-owned, municipal and private utilities, as well as users, now have clear rules in harmony with the new public policy.

The regulator has drawn up proposals for new legislation and the altering of existing legislation and the issuing of regulations and recommendations to contribute towards a clarification of the rules for the provision of these services. It has periodically carried out an analysis, balance and assessment of existing legislation, resulting in the identification of issues to be reviewed, particularly gaps, inconsistencies and outdated aspects, and the complementary aspects requiring new legislation.

2.4. Definition of the institutional framework

A new public policy should be supported by a well-structured administrative organisation. Portugal has established an appropriate institutional framework, with a clear assignment of responsibilities for the public entities involved, especially the regulatory authority for water and waste services and the environmental, water resources, waste management, public health, consumer protection and competition authorities, without prejudice to certain modifications and mergers of institutional solutions which have been verified over time.

This definition was absolutely fundamental and essential for a good performance by the sectors, since it has enabled the responsibilities of stakeholders to be specified, along with clear rules of operation and the articulation between the close and complementary sectors mentioned above, without overlaps or important gaps.

Currently, the institutional framework in Portugal includes essentially the regulator (Water and Waste Services Regulation Authority - ERSAR), the environmental and water resources authority (Portuguese

Environment Agency – APA), the public health authority (Directorate General for Health – DGS), the consumer protection authority (Directorate General for the Consumer – DGC) and the competition authority (Competition Authority – AdC). Almost all of them are on the advisory board of ERSAR, thus increasing their overall intervention efficiency in helping to clarify responsibilities and links between them.

The creation of the regulator was certainly an important step, and this is described in greater detail in Annexes B and C.

2.5. Definition of the governance models for the services

Portugal decided on the governance models that could be used in the sectors, in accordance with current political options. They are direct management, delegated management and concessions for water and waste services, with provision by State-owned, municipal and private entities. Currently there are around 80% of cases involving direct management, 10% of cases involving delegated management and another 10% involving concessions. In 15% of these situations private management is involved.

The introduction of these different models of governance, all of them having clear cases of success, has meant that the option adopted varies from municipality to municipality and from region to region and there is a comparative analysis of their performance and thus the stimulus to improve services.

The regulator ensures the legal and contractual monitoring of the utilities throughout their life-cycle, whatever their model of governance, particularly through analysing the tendering and contracting processes, contract modifications, contract terminations, and reconfigurations and mergers of systems, accompanying the carrying out of contracts and intervening where necessary in reconciliation activities between parties.

2.6. Definition of the access targets and the quality of service goals

Portugal has set access targets and quality of service goals suitable for the specific situation of the country, namely, the population that must have available public water and waste services, and quality of service objectives.

The way in which utilities have implemented these goals has evolved over time. The concept of good quality service has historically tended to pass through three stages, namely the quantity stage, in which the main task was to satisfy the basic quantitative needs of the population, the quality stage, where the water quality and environmental objectives were joined to the previous stage, and the excellence stage, which seeks to add to the quality stage the strand of sustainable development in social, economic and environmental terms.

The quality of service where specified and has been continually monitored with regard to the goals specified, based on indicators relating to the user interface, the sustainability of utilities and environmental sustainability.

The regulator has ensured the regulation of the quality of service provided to users by the utilities, assessing their performance and comparing the utilities among themselves, through the application of a suitable selection of performance indicators, so as to promote effectiveness and efficiency, that is to say, an improvement in their levels of service. It has also ensured drinking water quality regulation, assessing the quality of water supplied to the users, comparing the utilities among themselves and monitoring any non-compliances in real-time.

2.7. Definition of the tariff and tax policy

Portugal has set out a tariff policy for public water and waste services with the goal of promoting a gradual trend towards cost recovery, consistent with the economic capacity of the population. Although somewhat distant from a full application, the recovery of the costs of these services has gradually been carried out through tariffs paid by the users, supplemented where possible through transfers of European funding and, if necessary, through rates from the levying of taxes.

At the same time an attempt has been made to introduce tax instruments which encourage desirable behaviour, for example the rational use of water as a primary material or as final destination, through a water resources usage tax, or as final destination suitable for solid waste, through a waste destination tax.

The country has thus sought to evolve from a situation of low tariffs to the gradual full recovery of costs and, although the process is not yet fully completed, certain successes have been achieved. Nowadays there is a definition of principles and rules to follow and the regulatory instruments for a gradual change from the current situation of unsustainability for many utilities.

The regulator has ensured gradually the economic regulation of the utilities, thus promoting the regulation of prices to ensure efficient tariffs which are socially acceptable to users without prejudice to the necessary economic and financial sustainability of the utilities, within an environment of efficiency and effectiveness in the provision of their service.

2.8. Provision and management of financial resources

To implement a new public policy and carry out coverage of services goals in Portugal it was necessary to ensure the availability of important financial resources, both national resources and those originating from European funds. Indeed, in the last twenty years there has been and continues to be a significantly higher level of investment in infrastructure for water and waste, about 10 000 million euro.

It was also necessary to create capacity for the efficient management of these so important financial resources, knowing how to apply the resources where they could provide greater profits and benefits for society.

The regulator has contributed by supporting the government in defining suitable criteria for the awarding of this financing and the on-going monitoring of the results obtained.

2.9. Construction of the infrastructure

Portugal has throughout this period promoted the construction of the infrastructure for the provision of public water and waste services, using appropriate technology, naturally implying substantial costs, both in terms of initial investments and at the operational level.

In its supplying of water Portugal currently has 300 surface water abstractions and 5700 groundwater abstractions, 230 treatment plants, 2400 pumping stations, 8400 tanks and 100 000 km of distribution networks. Using this infrastructure, 276 utilities, with 9300 staff, abstract every year 850 million m^3 of water, using 600 million kWh of energy, and provide a service to 9.5 million individuals in Portugal.

For waste water there are 50 000 km of collectors, 4400 pumping stations, 2500 treatment plants, 1800 collective septic tanks and 26 submarine outfalls. Using this infrastructure, 284 utilities, with 5700 staff, using 340 million kWh of energy, provide a service to 8 million individuals in Portugal, collecting and treating 1000 million m^3 of waste water every year.

For solid waste management there are 383 073 solid waste collection containers, 66 173 eco-point bins, 2391 waste collection vehicles, 13 organic recovery plants, 2 incinerators and 34 landfills. Based on this infrastructure, 283 utilities, with 12 150 staff, provide a service to around 10 million individuals in Portugal, collecting every year almost 4.6 million tonnes of waste, of which 3.8 million tonnes come from unsorted waste collection and 800 thousand tonnes come from sorted collection. Each year 400 thousand tonnes of waste are recycled, 300 thousand tonnes of waste undergo organic recovery, for example composting, 1 million tonnes of waste are incinerated and 3.1 million tonnes are still deposited in landfill.

This infrastructure has enabled an enormous increase in the level of compliance with European legislation in this matter.

The regulator has contributed by monitoring and even approving new investments by utilities within the scope of economic regulation.

2.10. Improving the structural and operational efficiency

Portugal has been promoting the improvement of the structural efficiency of the public water and waste services and the efficiency of the operation by the utilities, to reduce costs to users and to society in general.

In improving the structural efficiency of the sectors, an optimised territorial organisation was specified for the management of these services, making use of economies of scale at the regional level. The utilities are therefore encouraged to promote, as much as possible, the physical integration and aggregation of the systems at a technically and economically appropriate scale, promoting the implementation of joint solutions with similar entities. In fact, nowadays there are clear benefits in the bulk service resulting from the physical aggregation of systems, not only in economic terms but also in terms of the quality of the services provided.

Another important measure is the promotion of economies of scope, joining the services of water supply to those for waste water. Process economies have also been assessed and discussed, with the aim of providing vertical integration of bulk systems (water production, treatment of waste water and solid waste) with retail systems (water distribution, collection of waste water and collection of solid waste).

In improving the efficiency of the operation of the utilities, the utilities should seek to adopt, given existing legislation, the most advisable organisation-type, particularly at the staffing level, functional content, information circuits, administrative procedures, planning, budget, control and measures which will tend to ensure quality assurance.

The regulator has contributed through the structural regulation of the sectors, which involves collaboration towards the improved organisation of the sectors, the clarification of its rules of functioning, the regular drawing up and dissemination of information on the sectors and for capacity building and innovation in the sectors, as well as the behavioural regulation of the utilities, which consists of their legal and contractual monitoring throughout the life-cycle, economic regulation, quality of service, drinking water quality and the user interface.

2.11. Human resource capacity building

A major effort regarding human resources capacity building has been carried out in Portugal in terms of quantity and competencies, as this is an essential factor to guarantee the general quality of the sectors.

There has been a strengthening of technological and traditional technical courses to overcome any shortages of personnel with relevant academic qualifications to carry out the existing roles in the sectors. There are training and updating courses for specific human resources in the sectors, involving training activities for managers and technicians at various levels.

There has been a considerable increase in technical publications and various other educational tools, of a practical character, covering all areas of the sectors, aimed at the different professional levels involved.

For twenty years Portugal did not have sufficient human resources in the sectors and it mostly had insufficient capacity building. The country currently has around 27,500 well-prepared professionals, with solid training in planning, design, financing, construction and operation of water and waste services, with recourse to advanced technology.

The regulator has provided technical support to the utilities through the production of technical publications in partnership with knowledge centres, the direct and indirect promotion of seminars and conferences, support for training events by third parties. In this way it can contribute to an improvement in the technical capacity building of the utilities.

2.12. Promotion of research and development

Portugal has promoted research and development in areas associated with public water and waste services, creating endogenous knowledge.

A research and development program for the sectors exists, covering pre-normative applied research projects, for the development and strengthening of the technological infrastructure on which they can base these projects. It has been possible to gradually promote innovation and technical support to the utilities, with research centres moving closer to industry, and ensuring greater national economy in terms of knowledge and technologies.

The Portuguese research centres have increased their participation in international research and development projects, particularly European ones, and in some areas even heading such initiatives.

The regulator has supported innovation and the promotion of research and development in the sectors, in partnership with the research centres.

2.13. Development of the economic sector

Portugal has sought to improve the development of the economic sector, taking advantage of the implementation of strategies to develop water and waste services, which has created exceptional conditions to promote the development of national knowledge and hence strengthen the capacity of the business sector in the domestic and international market, generating new activities with the creation of employment and wealth.

The regulator has provided technical support to the sectors and particularly the utilities through the production of technical publications in partnership with knowledge centres, the direct and indirect promotion of seminars and conferences, support for training events by third parties. In this way it can also encourage the consolidation of the national business sector.

2.14. Introduction of competition

In the case of natural or legal monopolies, and where there is therefore no competition in the market, Portugal has promoted virtual competition, through benchmarking between utilities and, in the case of private involvement, competition in the market, for example through tender procedures for the allocation of delegations, concessions and the provision of services.

Indeed, the introduction of different models of governance in Portugal has enabled competition to increase, as well as a comparative analysis of their performance and thus an on-going stimulus to improve services.

This competition encourages innovation and technical progress and, therefore, increases the efficiency and quality of the provision of these services.

2.15. Protection, awareness and involvement of the users

In Portugal the tools for the protection of users have been promoted, especially for the most economically disadvantaged, as well as awareness and involvement with regard to public water and waste services.

The users now have rights reinforced through specific legislation and increasingly involvement about water and waste services, particularly regarding physical and economic access to services, their quality, the quality of drinking water, information about services, complaints about services and their participation in decisions. Complaints have become a powerful instrument for the defence of users.

The involvement of the population in the decision-making processes has gradually been promoted through environmental education activities, for example regarding the efficient use of water and reduction in waste production and separation, and with the growing availability of information, has allowed users to make more well-founded decision regarding their preferences, particularly regarding the levels of coverage to be attained and the quality of services to be provided, with regard to what they are willing to pay.

The regulator continually ensures compliance by the utilities with consumer protection legislation and, in particular, undertakes an analysis of any complaints and promotes their resolution between users and the service provision utilities. It has also fostered the participation of the users of the services, creating advisory mechanisms and disseminating information. The users are naturally represented on the regulator's advisory board.

2.16. Provision of information

A very complete information system has been established in Portugal for the water and waste services, making use of reliable information, both to support the definition of public policies and business strategies and to evaluate the service that is actually provided to society, so as to be able to convey an overview of the sectors in a reliable and regularly updated manner.

Information is disseminated on two levels. The first is at an essentially national level, which is most useful for defining policies and development strategies, and a second essentially at the level of the utility, mostly useful for the operation of the systems. It is also made available to users in a more accessible format.

The regulator has the responsibility to regularly collect, validate, file and process information about the situation of the sectors, following a standardised format, which is sufficiently complete and easy to interpret.

3. RESULTS OF THE IMPLEMENTATION OF THIS POLICY

3.1. Overview

It is now summarized the results of the implementation of this policy in mainland Portugal during the last two decades (1993–2012), not only regarding the accessibility of the drinking water supply, waste water management and solid waste management service and the quality of drinking water, but also regarding the impact on environmental quality and also on public health.

3.2. Evolution of the public water supply services

In 1993 only 81% of the dwellings in mainland Portugal were covered by the public water supply service. Currently, 95% of dwellings are covered by the service, which means that the coverage target envisaged in the strategic plan for the drinking water supply and waste water management (PEAASAR) has been attained. The remaining 5% of dwellings are served by the individual solutions, such as individual holes or wells. There are, however, significant differences at the regional level between urban areas and rural areas, and it can be seen that around 99% of dwellings in urban areas have access to public water supply services, while in the rural areas access is less, in the order of 90%.

It can be noted that the public water supply service has significantly evolved and attained its global target. Naturally there continues to be needed a certain amount of investment with the aim of resolving localised problems and with concerns regarding patrimonial management, within a cost benefit perspective.

Figure A.1 shows the positive evolution of the coverage level of the drinking water supply service in the last two decades:

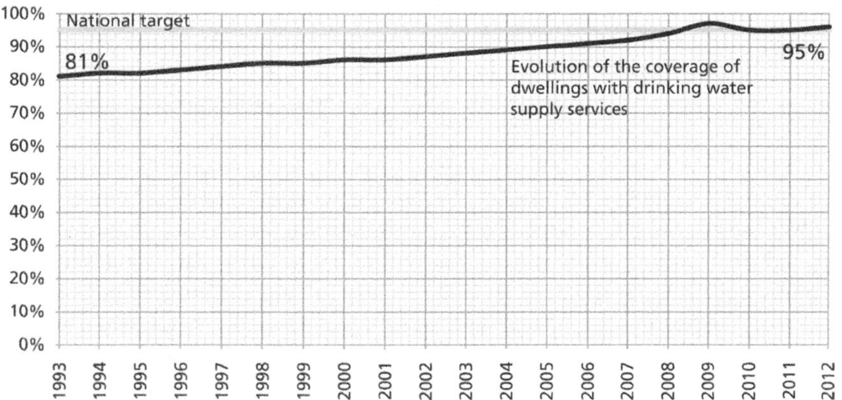

Figure A.1 Evolution of the physical accessibility of the drinking water supply service. (*Sources*: Entidade Reguladora dos Serviços de Águas e Resíduos – ERSAR, and Agência Portuguesa do Ambiente – APA).

As for the water quality, in 1993 only 50% of the dwellings in mainland Portugal were provided with safe water, in accordance with national and European legislation. These public services currently ensure a high quality level of distributed water, with nearly 99% complying with legislation, with the remaining non-compliant situations being subject to immediate corrective intervention.

It can be noted that the quality of water has evolved in an extraordinary manner. The target envisaged in the strategic plan for the drinking water supply and waste water management (PEAASAR) has practically been detained. The situation now is being one of maintaining, updating and perfecting existing mechanisms, from a cost benefit perspective. This is a remarkable example of a successful strategy, with a very positive impact on public health, on the reduction of diseases and deaths and the reduction of days absent from work.

Figure A.2 shows the positive evolution of the quality of water supplied in the last two decades:

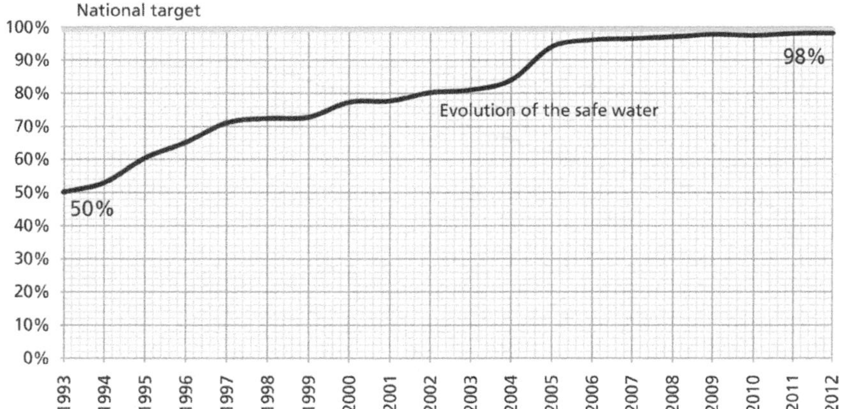

Figure A.2 Evolution of the quality of drinking water supplied. (*Source*: Entidade Reguladora dos Serviços de Águas e Resíduos – ERSAR).

3.3. Evolution of waste water services

In 1993 only 60% of the dwellings in mainland Portugal were covered by public waste water collection services. Currently 81% of dwellings are covered by these services. But what is intended is to ensure not only the collection of waste water and also it suitable treatment before its discharge into the environment, and there the situation was clearly worse. In 1993 only 28% of the dwellings in mainland Portugal were covered by public waste water services involving collection and treatment. Currently 79% of dwellings are covered by these services.

This means that there has been a major evolution. However, the goal of 90% envisaged in the strategic plan for the drinking water supply and waste water management (PEAASAR) has not yet been attained. The remaining dwellings (19%) are serviced by private solutions, such as septic tanks.

The geographical distribution of the physical accessibility of the waste water service shows that the differences are quite significant at regional level. There are greater differences between urban areas and rural areas, given that, despite municipalities with an urban characteristic having an access percentage of 95% for dwellings with a waste water management service, above that established as the national target (90%), in the case of more rural areas this is still situated at around 70%. These disparities also result from the fact that it can be seen that in the rural areas these public systems often have excessively high unit costs, with the option for individual solutions being preferable.

It can be seen that the waste water service still needs to be significantly improved, with investment continuing in a rational manner, guided by environmental, public health, and asset management objectives, within a cost benefit perspective. It is also necessary to improve the operational aspect. Indeed, 97.5% of the requested analyses for discharged waste water are carried out with regard to those envisaged in discharge licences or in legislation, and compliance with discharge parameters is 90%, so there is potential for improvement.

Figure A.3 shows the positive evolution of the waste water management service in the last two decades:

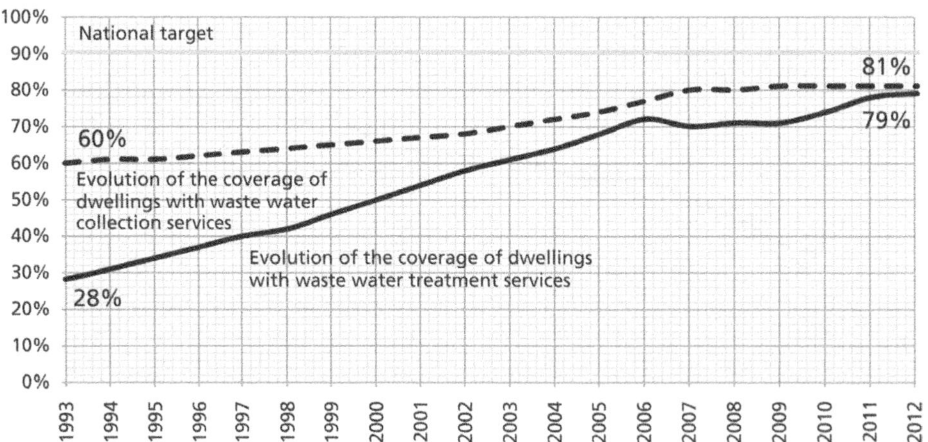

Figure A.3 Evolution of the physical accessibility to the waste water management service. (*Sources*: Entidade Reguladora dos Serviços de Águas e Resíduos – ERSAR, and Agência Portuguesa do Ambiente – APA).

3.4. Evolution of solid waste management services

In 1993 about 94% of the dwellings of mainland Portugal were covered by the solid waste collection service. Currently, 100% of dwellings are covered by this service, which means that the coverage target envisaged in the Strategic Plan for Solid Waste Management Services (PERSU) has been attained.

But what is intended is to ensure not only the collection of waste and also it suitable valorisation or treatment instead of its dumping into the environment, and there the situation was clearly worse. In 1993 only 23% of the dwellings in mainland Portugal were covered by public waste services involving adequate destination, with 9% involving valorisation. Currently 100% of the dwellings are covered by public waste services involving adequate destination, with 46% involving valorisation.

It can be seen that the solid waste management service needs to continue its investment, mainly in terms of technological evolution, but in a rational manner, related to environmental demands, within a cost benefit perspective. There is also a need to increase prevention concerning the creation, re-use, recycling and organic and energy recovery of waste.

Figure A.4 shows the positive evolution of the coverage level of the solid waste management service in the last two decades:

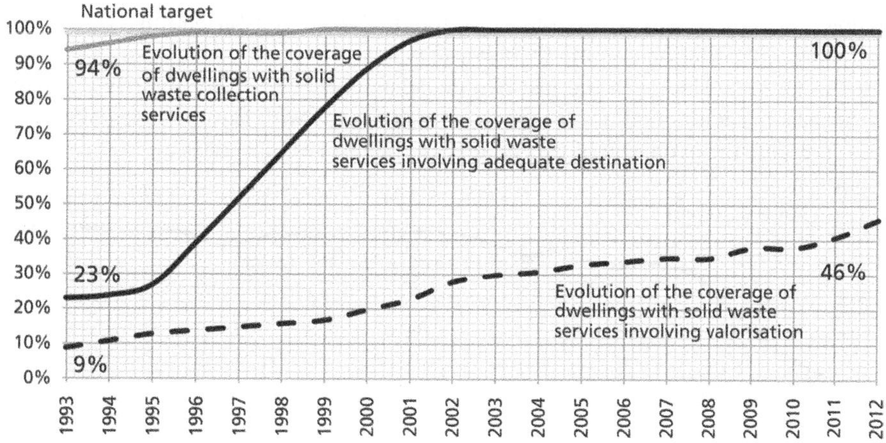

Figure A.4 Evolution of physical accessibility of the solid waste management service. (*Sources*: Entidade Reguladora dos Serviços de Águas e Resíduos – ERSAR, and Agência Portuguesa do Ambiente – APA).

3.5. Impact on the environmental quality

The water and waste services have a strong impact on environmental quality, particularly as regards the discharge of waste water into water resources.

A possible indicator to assess this impact is the quality of surface waters, insofar as it is obviously affected by any pollution coming from waste water, even diffuse pollution and other sources can contribute significantly.

It can be seen that its evolution in the last two decades has been very positive, going from 28% to 78% quality for surface waters good enough to be used for drinking water supply after treatment, in accordance with European legislation.

Figure A.5 represents the evolution of this quality in the last two decades:

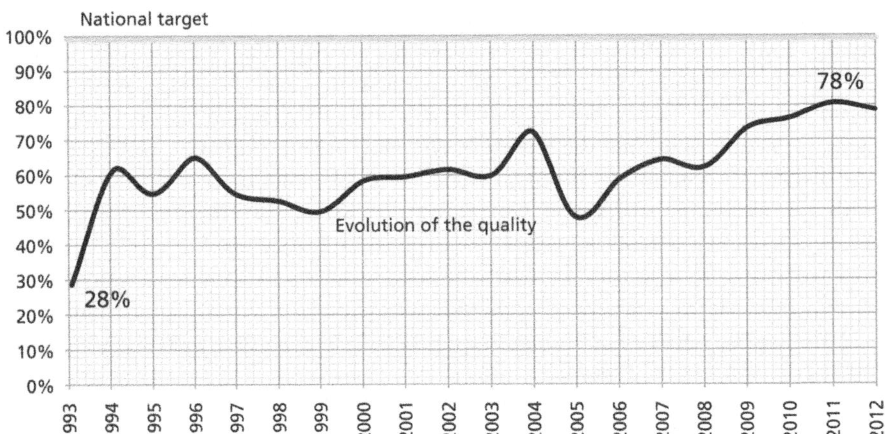

Figure A.5 Evolution of the quality of surface waters. (*Source*: Agência Portuguesa do Ambiente – APA).

Other possible indicator to assess this impact is the quality of coastal bathing waters, insofar as it is heavily impacted by any pollution coming from waste water.

It can be seen that its evolution in the last two decades has been extraordinarily positive, going from 58% to 100% for beaches with good water quality, in accordance with European legislation.

Figure A.6 represents the evolution of this quality in the last two decades:

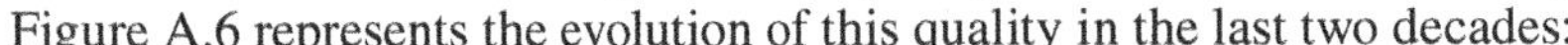

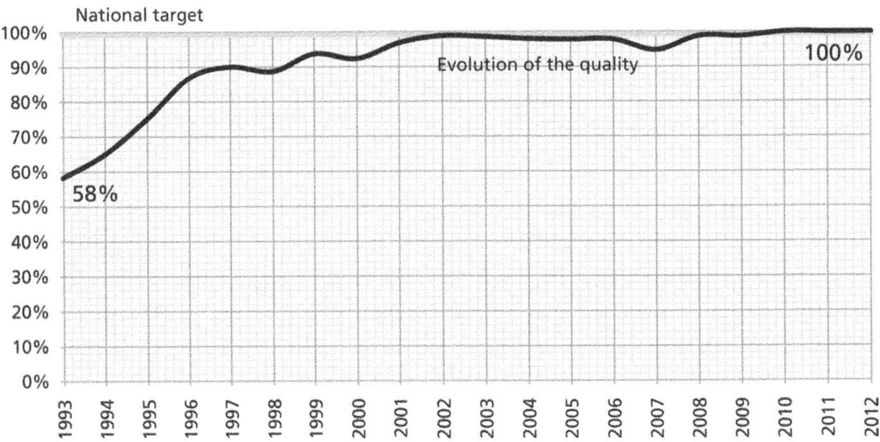

Figure A.6 Evolution of the quality of coastal bathing waters. (*Source*: Agência Portuguesa do Ambiente – APA).

Another possible indicator to assess the impact of the water and waste services on the environment is the quality of inland bathing waters, insofar as it is also more strongly affected by any pollution, particularly from the discharge of waste water.

It can be seen that its evolution in the last two decades has been extraordinarily positive, going from 17% to 95% for inland bathing waters with good water quality, in accordance with European legislation.

Figure A.7 represents the evolution of this quality in the last two decades:

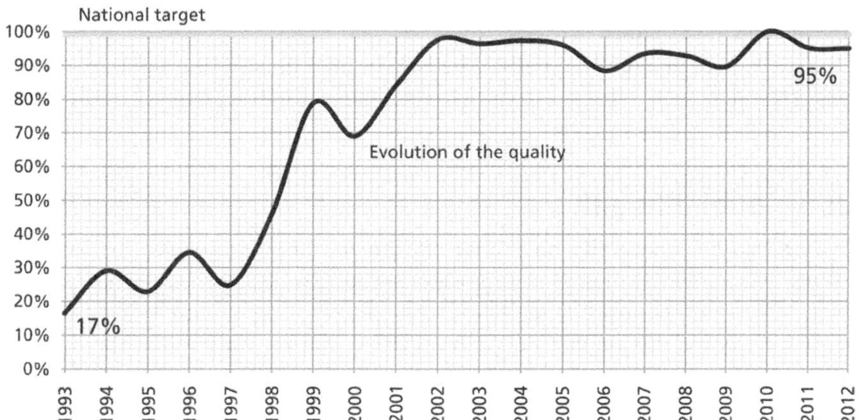

Figure A.7 Evolution of the quality of inland bathing waters. (*Source*: Agência Portuguesa do Ambiente – APA).

Other possible indicators to assess the impact of the water and waste services on the environment are the number of beaches with blue flag or classified as gold beaches. Blue flag is a distinction annually awarded by the Associação Bandeira Azul da Europa to coastal and inland beaches and marinas which comply with a set of requirements. These include various parameters, the first of which is the water quality, but also considering information and environmental education, conservation of the local environment, safety, services and infrastructure support. Additionally, gold beaches are those having an irreprehensible water quality verified in the last five years by the environmental protection association Quercus (process initiated in 2003). They both may, therefore, be considered as symbols guaranteeing the quality of the beaches.

It can be seen that, despite criteria having evolved from year to year, becoming ever more demanding, the number of beaches and marinas distinguished has been increasing in a very significant manner, going from 89 to 289 with a blue flag and from 87 to 293 with gold quality.

Figure A.8 represents the evolution of these indicators in the last two decades:

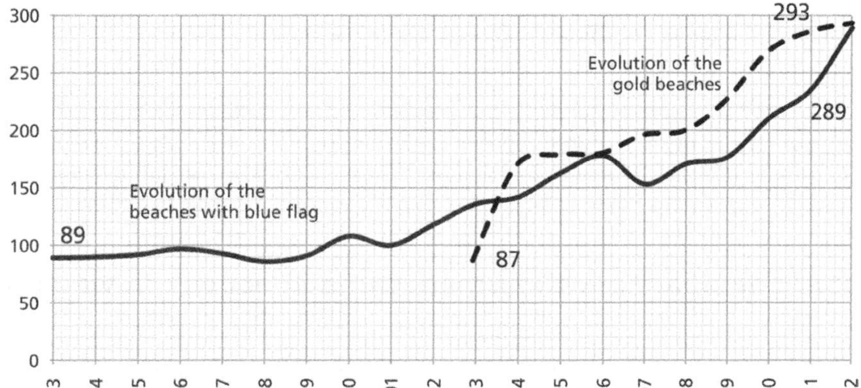

Figure A.8 Evolution of the number of beaches with blue flags and gold quality. (*Sources*: Associação Bandeira Azul da Europa – ABAE, and QUERCUS).

Summarizing, the improvement of water and waste services in Portugal had a strong impact on environmental quality, namely surface waters, coastal bathing waters, inland bathing waters and the number of blue flags and gold quality beaches.

In some of the cases it is apparently possible to identify the impact of drought years on water quality, provoking a temporary step back in the positive environmental evolution, namely in the dry periods 1994–1995 and 2004–2006, mainly the last one.

3.6. Impact on public health

The water and waste services naturally have a strong impact on public health, particularly through the quality of drinking water, which may transmit waterborne diseases, particularly cholera, typhoid and paratyphoid fever, other salmonelloses, shigellosis, leptospirosis, Legionnaire's disease and hepatitis A.

Using hepatitis A to assess this impact, it can be verified that its evolution in the last two decades has been very positive, going from 630 cases to 10 cases annually.

Figure A.9 represents the evolution of this disease in the last two decades:

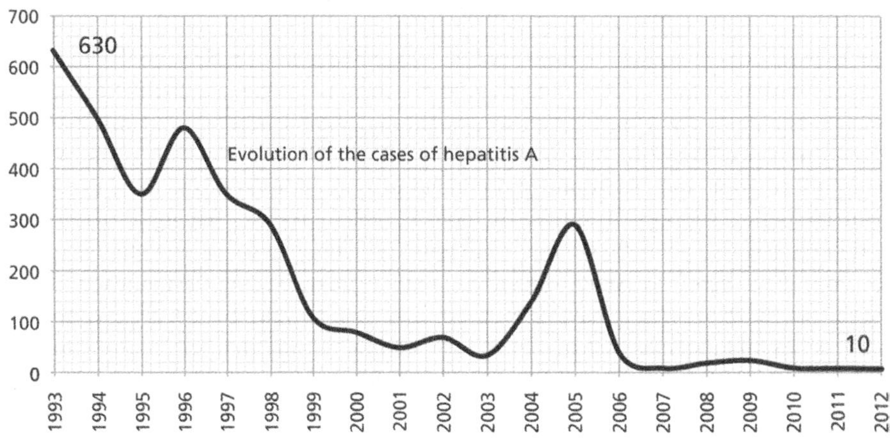

Figure A.9 Evolution of hepatitis A. (*Source*: Direção Geral de Saúde – DGS).

It is again apparently possible to identify the impact of drought years on hepatitis A, provoking a temporary step back in the positive public health evolution, namely in the dry periods 1994–1995 and 2004–2006, mainly the last one. This was not due to decrease of quality of public supply, which did not happened, but probably to the use of non-controlled water sources, like individual holes or wells.

3.7. Impact of compliance with human rights in access to water and sanitation

In 2010 the General Assembly of the United Nations declared access to drinking water and sanitation as an essential human right. This means that everybody should have suitable and safe access to water and sanitation, which can be carried out through traditional public systems (supply or sanitation networks), simplified public systems (e.g., collective septic tanks) or individual installations (e.g., individual septic tanks). The services and the installations should be physically accessible, and have suitable capacity, acceptable quality, be economically accessible and culturally adapted. Non-discriminatory access should be guaranteed to all, as well as the participation of citizens in the decision process along with the existence of monitoring and reporting. The implementation of these rights means, for Portugal and the member states, the obligation to respect, protect and ensure them, but this does not mean that the services are free.

In carrying out the level of implementation of equitable access to water and waste water, it is possible to state that Portugal has a clear public policy for the sectors, which in general is one that is very satisfactory, with a good compliance to the spirit of human rights in terms of access to water and waste water, as referred to above.

Around 98% of dwellings are satisfactorily equipped with water, bathroom, shower or bath facilities. 95% of households are served by the public water supply system, with the other situations resolved through individual solutions such as their own holes or wells. Drinking water quality is excellent, with a safe water index close to 99%. As far as waste water services are concerned, 79% of households are served by the public system including collection and treatment, with the other situations resolved through individual solutions such as septic tanks.

In addition, the country has been promoting several necessary measures, particularly legislative ones, to consolidate its human rights.

Citizens have the right to the provision of water services through the fixed networks when located less than 20 m from these. They have the right of supply of water services up to 5 days after the submission of their request and have the right to service continuity, which may only be interrupted for exceptional reasons.

There is the right that the criteria for defining the tariffs by the service holders should take into account the economic capacity of the populations, with a social tariff and a family tariff being recommended, along with the gradual doing away with an automatic charge for water service connections and a ban on a demand for a down payment. The tariff structure should include progressive blocks, which seek to ensure the service is economically accessible for a minimum volume of water supply, which is considered basic and essential for survival (the 1st block, which generally includes the first 5 cubic metres monthly).

There is the right to a social tariff, through the reduction of the tariffs of services for users whose household has a gross income lower than a certain value. This reduction may be carried out through the exemption or fixed tariffs and the application of variable tariffs for the first block of user consumption up to 15 cubic metres monthly.

They have the right to credible and easily interpreted information on the utility of the services, the respective tariffs, quality of service and the quality of drinking water.

They have the right to complain and have the complaint answered in a proper and timely manner, through independent intervention by the regulator as well as permanent consultation regarding its status, via the Internet.

An indicator of economic accessibility is now regularly calculated which measures the weight of the average charge for services of drinking water supply, waste water management and solid waste management on the average income available per household in the area served by each utility, and this should not go beyond 1%, per service provided, and should ideally be equal or less than 0.5%.

Suspension of the water supply service due to lack of payment is possible in Portugal, because experience has shown that the threat to cut-off supplies to non-payers is essential for sustainability. Many examples show that generally non-payment is not a matter of being unable to pay, but rather of being unwilling to pay. However, suspension of the water supply is only possible after sending the appropriate advance notice, cautioning the users of their interests in a service which is considered essential. This advance notice should be made in writing, with a minimum advance period of twenty days concerning the suspension of the service. It should also provide other information, particularly the reason for the suspension, that is, the identification of the amounts in debt, the means available to the user to avoid the suspension of the service and re-establishing the same, that is, locations, deadlines and means of payment, as well as information regarding the payment of the amounts demanded to avoid the suspension of the service or guarantee its re-establishment, notwithstanding that the user in general terms can assert their rights.

An area which still deserves greater attention is that of access by vulnerable and marginalised groups, such as the homeless, populations without fixed residence, and individuals with some form of physical incapacity.

Improvement of services in public areas can also be considered, especially in the areas of greater density, for example with more and modern sanitary and hygienic installations in central squares in public buildings, as well as spaces designed for occasional events, such as fairs, and short stays by nomadic and travelling populations.

4. CONCLUSION

From the analysis carried out, it can be seen that the new public policy instituted in 1993 for the water and waste services was implemented in a global and integrated manner, with greater stability over time. Success required continuity of government pursuing a well-defined public policy implemented through long-term planning.

It brought together institutional, governance, management, planning, technical, economic, legal and environmental, public health, social and ethical instruments to ensure the suitable provision of these services. In that new policy, regulation has played an important role.

As a consequence, a very positive evolution in the provision of these essential public services can be noted, particularly in terms of the evolution of the public services of drinking water supply, waste water management and solid waste management, and the impact on environmental quality and on public health and also the impact of compliance with human rights in access to water and sanitation.

Annex B

The evolution of water and waste services regulation in Portugal

1. INTRODUCTORY NOTE

Portugal may be compared with its European counterparts in the area of regulation of network services, having indeed been one of the first countries to create independent regulatory bodies in the 1990s, following an international trend for the liberalisation of traditional monopolist markets. This was the case with the telecommunications, electricity and gas sectors which, after liberalisation, needed regulatory bodies to supervise these recently created competitive markets. These bodies now have around two decades of existence and the experience has been, overall, positive, and these bodies have earned national and international prestige.

This has also been the case with the water and waste services from the beginning of the present century, in which Portugal was one of the pioneering European countries in setting up a regulatory body, with its intervention currently recognised in the country and abroad due to the rationality and robustness of its regulation model.

ERSAR founded together with the Italian Regulatory Authority for Electricity Gas and Water and other European regulators, in Milan, Italy, in April 2014 the *Network of European Water Service Regulators* (WAREG) in order to promote Europe-wide coordination between water service regulatory authorities and promote the best regulatory practices. ERSAR also organised in partnership with the International Water Association the *1st International Water Regulators' Forum,* as part of the *IWA World Water Congress & Exhibition* held in Lisbon, Portugal, in September 2014.

These positive experiences have created good conditions for the consolidation of independent, credible, strong and universal regulation in Portugal, defending the general interest and the uses of regulated services, without prejudice to safeguarding the economic viability of the utilities and their legitimate interests. The general acceptance by stakeholders of the sectors and consumer expectations regarding the regulators also constitute an element of legitimacy to reinforce the capacity building of these bodies to better carry out their role.

2. EVOLUTION OF THE REGULATION AUTHORITY

In Portugal, the setting up of a regulatory authority had from the outset been envisaged in the reform of the drinking water supply, waste water management and solid waste management, undertaken from 1993 onwards, with new public policies for these services.

Indeed, following the alteration of the law delimiting the sectors in which access to private capital was permitted, the approval in 1993 of Decree-Law No. 379/93, of 5 November, which enshrined the legal system for the management and operation of systems for the purpose of water and waste services, opened the way for the approval of other structural laws for regulation of the sectors.

The first attempt to regulate these services was taken in 1995 with the setting up of the Concessions Monitoring Committee established by Order No. 38/MARN/95, of 26 August, with powers delegated for the State regarding the monitoring and accompanying of the concessionaire and authorisation for its acts, which drew up opinions on the investment plans of the concessionaire companies of multi-municipal systems providing bulk services to various municipalities and with regard to the tariff systems proposed by them.

Shortly afterwards, there was the setting up of the National Observatory for Municipal and Multi-municipal Systems, through the approval of Decree-Law No. 147/95, of 21st June, which attributed it with functions involving the analysis of tender bidding processes and consideration of the minutes of concession contracts, the collection of information on the quality of service and water quality and the dissemination of this, as well as the drawing up of recommendations to licensors and concessionaires. This body, however, was never set up, and was made extinct three years later.

The Institute for the Regulation of Water and Solid Waste (IRAR) was established in 1997 by Decree-Law No. 230/97, of 30 August, with the duties of the water and waste services regulator in Portugal, as a public institute endowed with administrative and financial autonomy but subject to the responsibility of the minister for the environment, distinct, therefore, from the model of the independent regulatory body existing at that date in other network services. Its powers were limited and it acted more using its powers of influence through constructive interaction with the other sectors' stakeholders.

Its bylaw was approved in 1998 by Decree-Law No. 362/98, of 18 November, with the alterations introduced by Decree-Law No. 151/2002, of 23 May. According to this bylaw, as far as municipal concessions were concerned, IRAR would issue an advice on the tendering process but would not intervene in the process of establishing the tariff, which constituted a determining element in the choice of the concession holder, with this element being regulated by the contract, as well as its updating. On the other hand, IRAR would also issue an advice on the tariffs for multi-municipal concession systems, annually proposed to the State by the concession holder. However, as far as the management models not operated under a concession, no regulatory activity was foreseen.

From 2003, IRAR was given the responsibility of the national authority for drinking water quality, under the scope of Decree-Law No. 243/2001, of 5 September, later revised by Decree-Law No. 306/2007, of 27 August, resulting from the transposition into national law of Council Directive 98/83/EC of 3 November. As well as this duty there was also a vast number of functions to be covered, with reinforced powers, of note being the approval of the water quality control programmes presented annually by the utilities, the appraisal, validation and supervision of the methods used and the laboratories undertaking analyses which provide this service to the utilities, the monitoring, inspection and possible sanctioning of the utilities and the drawing up of annual reports on the situation concerning compliance with legislation.

After some years of activity, Resolution of the Council of Ministers No. 72/2004, of 16 July, concerning the restructuring of the water sectors, would consider regulation an essential part for the development of the water sectors, within a preferentially competitive market. This was the real indicator of the changeover in Portugal from the infrastructure stage to a stage characterised by the safety and quality of the service provided. It was thus considered necessary to carry out a redefinition of the regulatory model, particularly in terms of goals, administrative nature, organic and functional independence, neutrality and universality of the regulated utilities. However, despite this need being identified, and notwithstanding the Public Institutes Restructuring Committee (CRIP), which concluded its work with the presentation of a final report on 4 July 2006, and, along the same lines, having proposed the possible transformation of IRAR into a special status public institution (independent administrative body), this did not take place.

The organic law of the Ministry for the Environment and Regional Planning (MAOTDR) approved through Decree-Law No. 207/2006, of 27 October, that IRAR remained within the indirect administrative sphere of the State with its statute as a public institute, with this diploma envisaging its restructuring and alteration to be called the Water and Waste Services Regulation Authority (ERSAR).

The legal system for the local business sector, as expressed in Law No. 53-F/2006, of 29 December, extended ERSAR's regulation to municipal companies, supplemented by Law No. 2/2007, of 15 January, which approved the Law of Local Finances and which would attribute responsibility to the regulatory authority for the verification of provisions regarding the prices of the services provided by municipal or intermunicipal direct management bodies, including those provided in the form of municipalised or intermunicipalised services, and by municipal and intermunicipal companies. This extended, once again, its scope of intervention. Further within the spirit of regulating and planning the water and waste sectors, Decree-Law No. 194/2009, of 20 August, and Decree-Law No. 195/2009, of 20 August, were published, which established, respectively, the legal systems for the municipal services and multimunicipal services for the public drinking water supply, waste water management and solid waste management, which confirmed and intensified the powers of the regulator.

The legal regulation framework was reinforced with Decree-Law No. 277/2009, of 2 October, which approved o new bylaw of ERSAR. Regulation of the sectors was reinforced, leading to the extension of the scope of intervention to all the utilities of these services, independently of their management model, as well as a greater uniformity of the regulatory body's procedures with regard to all of them. Therefore, in accordance with this legal framework, the direct management of services involving State and municipal ownership (municipal, municipalised and intermunicipalised services) became subject to quality of service regulation. Economic regulation now covered all the municipal services managed by contract (concessions, delegations and partnerships) and municipal services managed without a contract (municipal and municipalised services), following a mechanism involving supervision of tariffs, in the first case, and tariff verification mechanisms through sampling, in the second. However, this diploma maintains the administrative nature of the regulatory body as a standard public institution, being governed by the legal system for public institutions, with certain derogations present in the aforementioned diploma.

In 2010, the Water and Waste Services Regulator for the Azores (ERSARA) was set up in the Autonomous Region of the Azores by Regional Legislative Decree No. 8/2010/A, of 5 March.

In 2011, the plan of Portugal's 19th Constitutional Government took on the goal of transforming the bodies with regulatory functions dependent on government, where the regulated markets justified this due to their importance, into independent administrative authorities, reintegrating the others into the traditional administrative service.

In the same way, the Memorandum of Understanding on Specific Economic Policy Conditionality, of 27 June 2012, established by the Portuguese State, the European Commission, the European Central Bank and the International Monetary Fund, enshrined the goal of ensuring that the national regulatory authorities would have the independence and the resources necessary to carry out their responsibilities.

It was considered that ensuring a balanced market is not possible if one of the parties, the State, takes on conflicting roles, being both an active agent and the entity which establishes the rules regarding the functioning of the sectors. This duality normally has the consequence of imbalance in the market function, to the detriment of appropriate performance.

It is therefore important to emphasize the autonomy of the regulator with regard to the executive power, granting it this status and the means which enable it to defend what is essential, the general interest and the interests of the users of the regulated services, without prejudice to safeguarding the economic viability of the utilities and their legitimate interests. Effective regulation was used to establish an incentive to increase the efficiency and effectiveness of utilities, thus avoiding the risk of the prevalence of these with regard to the users and the subsequent risk of these users receiving services of lower quality and at a higher price.

In maximising the level of intervention of the regulator, within the scope of good regulation practices, it intends to contribute towards the reduction of risk in the sectors and suitable capacity building of funding, along with the protection of the interests of the users, thus guaranteeing long-term sustainable options. It is equally intended to ensure the safeguarding of the rights of the utilities and, where applicable, conditions of equality and transparency in access to the carrying out of the water and waste services activity and the respective contractual relationship, as well as consolidating an effective public right to general information regarding the sectors and regarding each of the utilities.

As such in 2013, through Law No. 67/2013, of 28 August, a framework-law was approved for independent administrative bodies with functions involving the regulation of the economic activity of the private, public and cooperative sectors, which standardised the legal framework of operation of these bodies. This also envisaged the alteration of the legal nature of ERSAR, placing it on the same level as other independent regulatory bodies.

Following this, and with a view to achieving all these goals, the Portuguese Parliament in 2014 approved and published Law No. 10/2014, of 6 March, containing the current bylaw of ERSAR. This diploma seeks to recognise and accentuate its autonomy with regard to the executive power, granting it this status and the means to defend the general interest and the interests of the users of the regulated services, as well as safeguard the economic viability of the utilities and their legitimate interests.

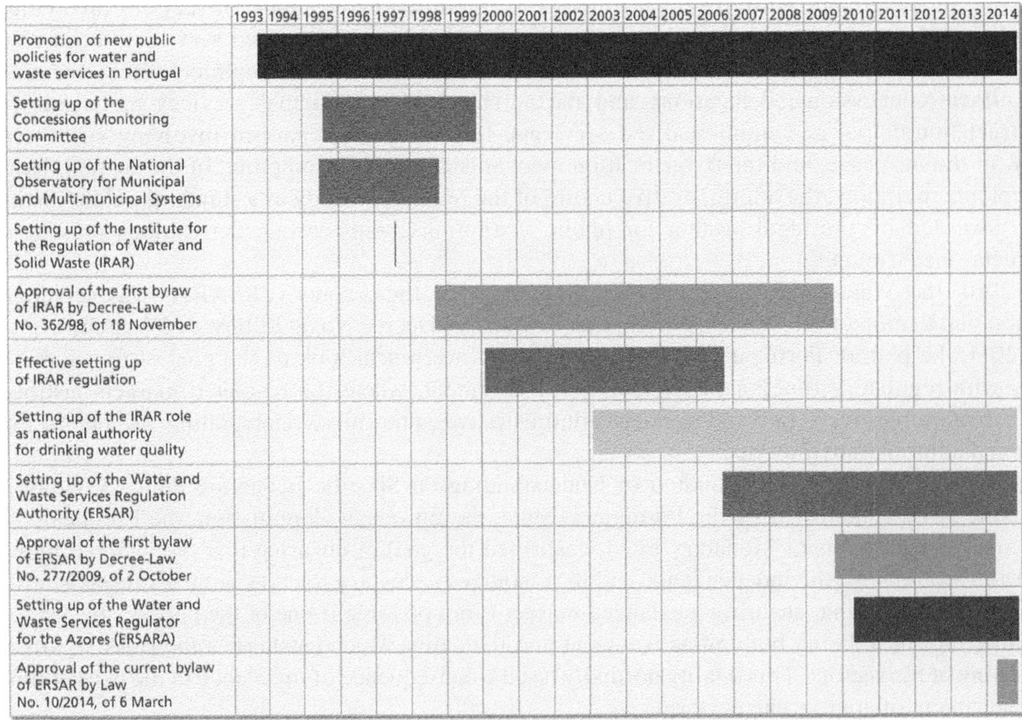

Figure B.1 Evolution of water and waste services regulation.

The aforementioned diploma reinforced the independence of the regulatory authority in carrying out its respective functions, particularly through the reduction of the powers to government and intervenes in its regulation and supervision under the terms of the law and its bylaw. It also alter the statute of the members of its board of administration concerning their appointment process, period of mandates, guarantees of immovability and the rules for the termination of mandates. The powers of authority and the sanctuary and legislative powers of the regulatory body were reinforced, to increase its capacity to act within the regulated sectors, through the attribution and strengthening of essential instruments for the regulatory and supervisory activity of the entities providing water and waste services.

The evolution over time of water and waste services regulation is shown in Figure B.1.

3. BOARD OF DIRECTORS

The creation of IRAR, and by Resolution No. 117/99 (2nd Series) of the Council of Ministers of 29 July, published in the Official Gazette, Series II, of 17 August, following a proposal from the then Minister of the Environment, the members of the first Board of Directors of IRAR, Pedro Cunha Serra, António Teixeira Cardoso and Adozinda Pinto were appointed, and took office on 14 September of 1999. Pedro Serra, at his request, left his post as Chairperson of IRAR on 14 March, 2001, with the management of the Institute being ensured by two board members.

By Resolution No. 26/2003 (2nd Series) of the Council of Ministers of 6 March, published in the Official Gazette, Series II, of 24 March, which came into effect from 18 March, the new Board of Directors of the Institute was appointed, consisting of Jaime Melo Baptista, Dulce Álvaro Pássaro and Rui Ferreira dos Santos, whose terms of office ended in March 2006. They were reappointed to carry out their duties performed in the Board of Directors of the Institute, by Resolution No. 60/2006 (2nd Series) of the Council of Ministers, of 8 June, published in the Official Gazette, Series II, of 26 June, and with effect from 18 March. The mandates of Jaime Melo Baptista and Dulce Álvaro Pássaro were renewed for a further period of three years, by Resolution No. 13/2009 published in Series II of Official Gazette, No. 109, of 5 June, 2009, with effect from 19 March, 2009.

At his request, Rui Ferreira dos Santos stepped down from the post of Member of the Board of the regulatory authority. In his place, João Simão Pires was appointed, in accordance with Resolution No. 100/2006 of the Council of Ministers of 26 October, published in the Official Gazette, 2nd Series, of 30 November, with effect from 2 November 2006.

Following the departure of Dulce Álvaro Pássaro on 26 October 2009, Fernanda Maçãs was appointed to the post of Member of the Board of the regulatory authority, by Resolution of the Council of Ministers No. 6/2010, published in the II Series of the Official Gazette, No. 47, of 9 March, with effect from 1 March.

With the mandate of João Simão Pires having been completed, Carlos Lopes Pereira was appointed Member of the Board of Directors of the regulatory authority by Resolution of the Council of Ministers No. 25/2010, of 24 June, published in the II Series of the Official Gazette, No. 137, with effect from 15 July, 2010.

Fernanda Maçãs resigned from the aforementioned post, with effect from 1 January, 2012. Therefore, the current Board of Directors of this Regulatory Authority consists of Jaime Melo Baptista and Carlos Lopes Pereira.

With the approval by Parliament of the current bylaw of ERSAR through the publication of Law No. 10/2014, of 6 March, the board of directors became the board of administration, the collegiate body responsible for the definition and implementation of its activity, as well as the management of the respective services, made up of a chairperson and two members, with one of them able to be designated as vice chairperson.

The evolution over time of the board of directors is shown in Figure B.2.

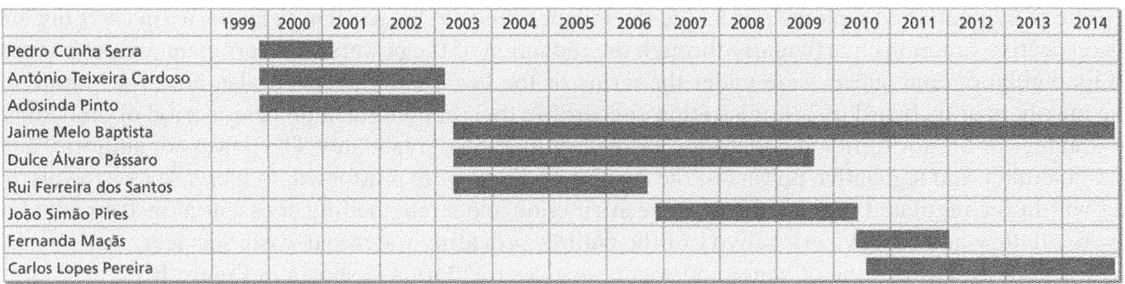

Figure B.2 Evolution of the board of directors.

4. ADVISORY BOARD

According to the bylaw, the Advisory Board is a consultative body to define the general lines of action of the regulatory body with the competence to issue advices concerning matters within its powers and those given to it by the Governing Board of Directors. It should, therefore, necessarily be heard regarding the plan and the annual activity report.

Upon the date of the setting up of IRAR, on 7 October 1999, Victor Martins was appointed to the post of Chairman of the Advisory Board of the regulatory authority, by order No. 20586/99 (2nd series) of the then Minister of the Environment, on 18 October, published in the Official Gazette, II Series, on 29 October, under the terms and pursuant to the provisions of Article 16 of the Statute annexed to Decree-Law No. 362/98, of 18 November, whose term ended in October 2002.

The activity of this body was suspended until the appointment of Francisco de Albuquerque Veloso as chairman of the Advisory Board, by order No. 16577/2003 (2nd Series) of the Ministry of Cities, Spatial Planning and the Environment, of 3 August, published in the Official Gazette, II Series, of 26 August.

Amilcar Augusto Contel Martins Theias was appointed as the new chairman of the Advisory Board through order by the Minister of the Environment, Land Use and Regional Development on 10 November, 2005, making it possible to reactivate this important body to support the functioning of the regulatory authority which brings together representatives of the different stakeholders and which has been very active and very important in the regulatory activity. By order of the Minister of the Environment, Land Use and Regional Development, of 9 January, 2009, Amilcar Augusto Contel Martins Theias was appointed again Chairman of the Advisory Board of the regulatory authority, with effect from 10 November 2008.

The evolution over time of the presidency of the advisory board is shown in Figure B.3.

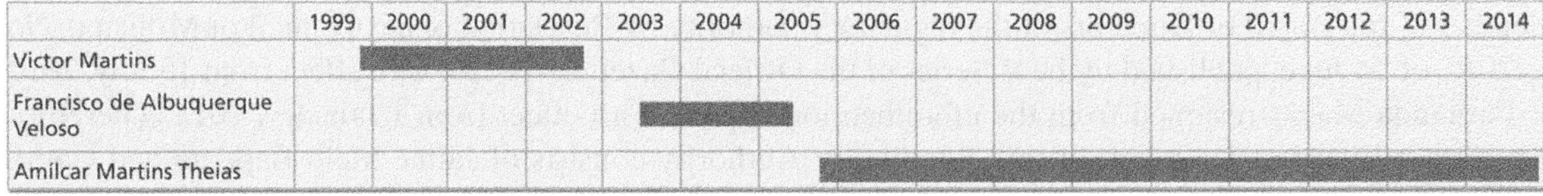

Figure B.3 Evolution of the presidency of the advisory board.

5. SOLE AUDITOR

According to the Organic Law, the Sole Auditor has the duty to monitor and control financial management, assess and issue an opinion on the budget and the annual activity report and the annual accounts of the regulatory authority, to supervise the proper implementation of accountancy practice and compliance with the applicable provisions with regard to budget, accounting and treasury matters, informing the Board of Directors of any anomaly that might be encountered, and rule on matters within its competence which are submitted to it by the Board of Directors. The Sole Auditor is responsible for monitoring the legality, regularity and sound financial and asset management of the regulatory authority.

Upon the date of the setting up of IRAR, on 7 October 1999, José Domingos Barão was appointed to the post of Sole Auditor of IRAR, by joint order No. 904/99 of the Ministers of Finance and the Environment, published in the Official Gazette, II Series, of 22 October, under and pursuant to the provisions of Article 19 of the Statute annexed to Decree-Law No. 362/98, of 18 November, whose term ended in October 2002, remaining in office until October 2003.

Maria do Rosário Líbano Monteiro was then appointed to the post of Sole Auditor of the regulatory authority, by Joint Order No. 969/2003, of 4 September 2003, of the Minister of State for Finance and the Minister for Cities, Land Use and the Environment, published in the Official Gazette, Series II, of 11 October. Her appointment as Sole Auditor was renewed with effect from 5 September, 2006, according to Joint Order No. 4620/2007 of 4 September 2006, of the Minister of State and of Finance and the Minister for the Environment, Land Use and Regional Development, published in the Official Gazette, Series II, on 14 March 2007. Joint order No. 3408/2010, published in the Official Gazette, Series II, on 24 February, renewed the appointment of Maria do Rosário Líbano Monteiro to the post of Sole Auditor for the regulatory authority, with effect from 5 September 2009.

The evolution over time of the sole auditor is shown in Figure B.4.

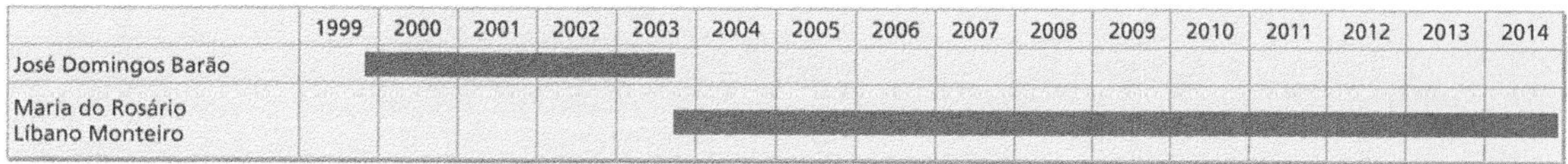

	1999	2000	2001	2002	2003	2004	2005	2006	2007	2008	2009	2010	2011	2012	2013	2014
José Domingos Barão																
Maria do Rosário Líbano Monteiro																

Figure B.4 Evolution of the sole auditor.

6. CONCLUSION

Analysing the evolution of water and waste services regulation in Portugal, it can be seen that the setting up of a regulatory authority for the water and waste services in Portugal resulted from a new public policy established in 1993, and that this was seen as an indispensable component for the promotion of this policy.

Six more years went past, however, until the start of its operation, which only took place at the end of 1999. The initial period involved certain fragility, and around a decade was necessary (1993–2003) for it to become more operationally effective.

It was also seen that, from that date, around another decade (2003–2014) of intensive regulatory activity was necessary, with mitigate powers, for it to become an independent administrative body, which is currently considered in many countries as the most suitable model, with reinforced functional and financial organisational independence, as well as the reinforcement of its powers of regulation, especially in terms of typical legal instruments (setting tariffs and binding instructions, etc.), as well as its regulatory competences, power to impose penalties, settlement of disputes and public disclosure of information.

This was thus a long but consistent and fruitful period of implementing a regulatory culture in the water and waste services, benefiting from reasonable stability both in terms of public policy for the sectors and in terms of the regulatory authority itself, regarding the management body, the advisory body and the supervisory body.

In the last decade it has thus been possible to conceive, develop and gradually implement an integrated regulatory model which includes structural regulation of the sectors, through contributing to its organisation, legislation, information and capacity building, and the behavioural regulation of the utilities, through legal and contractual, economic, quality of service, quality of drinking water and user interface regulation. This is the integrated regulatory approach (model RITA-ERSAR) which has been described in this book.

Annex C

The bylaw of the Water and Waste Services Regulation Authority (ERSAR)

In Portugal the bylaw of the Water and Waste Services Regulation Authority were approved by Law No. 10/2014 of the parliament, on 6 March, after intense debate with all the sectors' stakeholders, and which is reproduced below.

It represents a good balanced example of political decision-making regarding the nature, mission and the duties of the authority, its powers in terms of authority, sanctions and regulations, its organic structure and staffing, its independence and asset, budgetary and financial system.

CHAPTER I: General dispositions

Article 1: Nature, mission, jurisdiction and head office

1 – The Water and Waste Services Regulation Authority, hereinafter known as ERSAR, legal person in public law, is an independent administrative authority with regulation and supervision functions, endowed with management, administrative and financial autonomy, and its own assets. It is attached to the Ministry with the environmental remit.

2 – ERSAR's mission is the regulation and supervision of the public drinking water supply, waste water management and solid waste management services, abbreviated to water and waste services, including carrying out the functions of the national authority for the coordination and inspection of the drinking water quality.

3 – ERSAR has jurisdiction in national territory, without prejudice to the provisions of the political and administrative statutes of the autonomous regions.

4 – ERSAR has its head office in Lisbon, and may set up other branches or other forms of representation, whenever its board considers this suitable for carrying out of the duties of ERSAR.

Article 2: Legal regime and independence

1 – ERSAR is independent in carrying out its functions, under the terms envisaged in the framework law for regulatory authorities, approved by Law No. 67/2013, of 28 August, and in the present bylaw, not being subject to governmental supervision or guardianship within the scope of this activity.

2 – *ERSAR is governed by the provisions of international and European law, by the present statues, by internal regulations and provisions that are specifically applicable to it and, in matters of financial and asset management, in terms of what is not envisaged in them or what is seen as incompatible with these, through the norms applicable to State-owned companies.*

3 – *Under the terms of paragraph 1 and in carrying out public powers, in all that is not contrary to the provisions of the present bylaw and the diploma which approves them, the following are applicable to ERSAR:*

   a) The code of administrative procedure and any other norms and principles of a general nature regarding the administrative acts of the State;

   b) The laws of administrative litigation, when acts undertaken in carrying out public functions and contracts of an administrative nature are concerned;

4 – *The following are also applicable to ERSAR, specifically:*

   a) The public procurement framework;

   b) The liability regime of the State;

   c) The duties of information resulting from the State organisation information system (SIOE)

   d) The jurisdiction and financial control regimes of the court of auditors;

   e) The inspection and audit services system for the State.

Article 3: Principle of speciality

1 – *The legal capacity of ERSAR encompasses the possession of rights and obligations necessary to carry out its objective, exercising its powers within the scope of its respective duties, making use of its resources for the purposes with which it has been endowed.*

2 – *ERSAR may, whenever it is asked to do so or through its own initiative, provide technical support and consultancy to parliament and the government.*

Article 4: Regulated entities

1 – *Also subject to ERSAR's activity, within the scope of its duties and under the terms of the present bylaw, are all utilities working in the sectors referred to in paragraph 2 of Article 1, independently of State or municipal ownership of the respective systems and the management model adopted, namely:*

   a) Direct provision of service;

   b) Delegation of the service to a State-owned company in the sectors, from the local business sector, intermunicipal bodies or a company created in partnership with the State;

   c) Service concession.

2 – *Also subject to ERSAR's activity, under the terms of the law, are those in possession of the water and waste services, whenever the rights and obligations of the utility or the users are concerned, as well as the laboratories which carry out quality control on drinking water.*

3 – *Also subject to ERSAR's activity, are the parishes and user associations to which these services have been delegated and, for the purpose of the present bylaw, are made equivalent to utilities of systems with municipal ownership, under the model envisaged in subparagraph b) of paragraph No. 1.*

4 – *Also subject to ERSAR's activity, are any other bodies which have taken on responsibility for the management of services under the scope of the regulated sectors, independently of their public or private nature and the ownership which legitimises the carrying out of those activities, which, for the purposes of the present statues, are made equivalent to utilities of systems of State or*

municipal ownership under the models envisaged in subparagraph b) or c) of paragraph No. 1. according to each case and with any necessary adaptations.

5 – For the purposes of the provisions of the previous paragraph, indicators are set up with regard to the transfer of responsibility through the management of services carrying out investments remunerated in whole or in part by tariffs charged to users, the assumption of risk in terms of demand, the charging of services to users and the length of the contractual term.

6 – ERSAR also regulates any other bodies which, by law, remain governed by its activity, particularly bodies with private systems for the public supply of drinking water, under the terms of Decree-Law No. 306/2007, of 27 August, altered by Decree-Law No. 92/2010, of 26 July.

Article 5: Duties

1 – The general duties of ERSAR are to ensure the regulation and supervision of drinking water supply, waste water management and solid waste management services, promoting an increase in the efficiency and effectiveness of their delivery, considering the protection of the rights and interests of users, ensuring the existence of conditions which enable an economic and financial equilibrium to be obtained on the part of the activities of the regulated sectors carried out through a public service framework, as well as the carrying out of functions by the national authority for drinking water along with the all the water supply utilities.

2 – The duties of ERSAR for the structural regulation of the sectors are:
   a) Collaborate with parliament and with the government in formulating public policies and legislation regarding these regulated services;
   b) Contribute towards a rationalisation and resolution of dysfunctions regarding these regulated services and the organisation of the sectors, as well as monitoring and reporting on the implementation of its strategic plans;
   c) Contribute towards a clarification of the rules for the provision of these services through the issuing of regulations and recommendations, and monitor the application of those regulations and recommendations as well as legislation currently in force.

3 – The duties of ERSAR regarding the behavioural regulation of economic aspects are:
   a) Establishing the tariffs for the systems involving State ownership, as well as supervising other economic and financial aspects involving the utilities of systems with State ownership, particularly the issuing of opinions, proposals and recommendations, under the terms defined in legislation and applicable regulations;
   b) Regulating, assessing and auditing the establishing and application of tariffs in systems with municipal ownership, whatever their management model, under the terms defined in legislation and applicable regulations.
   c) Issue recommendations concerning the conformity of the municipal systems tariff with that established in tariff regulation and other applicable legislation, as well as inspecting and sanctioning any non-compliance;
   d) Issue, for the situations and terms envisaged in law, binding instructions regarding the tariffs to be practised by systems of municipal ownership which do not conform to the legal dispositions and regulations in force.
   e) Ensure the detailed invoicing by the utilities providing the services, within an identification framework made up of various parts which make up the final value of the invoice, so as to provide a breakdown for the final user of the different cost components regarding drinking water supply, waste water management, solid waste management and other activities.

4 – *The duties of ERSAR regarding behavioural regulation also include:*

    a) *Monitor compliance by the holders and managers with the provisions, regulations and applicable contracts, particularly in the stages involving the setting up, tendering, contracting, altering of contracts, reconfiguration and termination, thus ensuring public interest and legality;*

    b) *Ensure drinking water quality regulation with the water supply utilities, under the terms established in applicable legislation, promoting improvement in its quality and universality, through evaluating the performance of those utilities;*

    c) *Ensure the regulation of quality of service provided to users by utilities, promoting improvement in levels of service, assessing the performance of those utilities, comparing the utilities amongst themselves and rewarding best practices;*

    d) *Promote comparison and the public dissemination of utility activity, thus embodying a basic right of users to have access to information, and thus consolidating a culture of making concise and credible information available which is easily interpreted;*

    e) *Ensure and safeguard the rights and interests of users with regard to tariffs, services and quality of service and promote the resolution of conflicts with utilities concerning these aspects;*

    f) *Foster the participation of service users, creating advisory mechanisms and disseminating information;*

    g) *Be aware of user complaints and the conflicts involving utilities, and through analysing these, promote use of conciliation and arbitration services between the parties as a way of resolving conflicts and taking steps which it considers urgent and necessary;*

5 – *The specific duties of ERSAR also include the following supplementary regulatory activities:*

    a) *Coordinate and carry out the collection and dissemination of information regarding the services of public drinking water supply, waste water management, solid waste management and the respective utility holders, thus ensuring the right of access to information for all users;*

    b) *Promote research, innovation and the carrying out of studies on matters concerning its duties, thus contributing to the improvement of the technical ability of the utilities and other sectors' stakeholders.*

6 – *ERSAR shall also carry out any other functions attributed to it by law.*

Articles 6: Duties of collaboration and the provision of information

1 – *All bodies, public and private, should cooperate with ERSAR so that it can obtain requested information in order to carry out its duties.*

2 – *Without prejudice to other legally established periods, and for the purposes of the provisions of the former paragraph, ERSAR may establish a maximum period of 30 days for the regulated entities to send it any necessary information so that it can carry out its functions in a suitable manner.*

Article 7: Relations of cooperation and collaboration

1 – *ERSAR shall establish forms of cooperation, collaboration and association, within the scope of its duties, with other entities of public or private law, at a national or international level, when this has been shown to be necessary or important in the carrying out of its respective duties.*

2 – ERSAR, under the terms of specific legislation within the framework of its duties, collaborates with other national regulatory bodies, particularly with the competition authority and the national authority for waste regarding specific integrated waste stream systems.

3 – The collaboration referred to in the previous paragraph deals with strategic definition aspects, the licensing of utilities and the definition and revision of counter value, instantiated through proceedings to define the regulation of regulatory procedures.

CHAPTER II: Carrying out authority, sanctionary and legislative powers

Article 8: Equivalence

In the carrying out of its duties, ERSAR assumes the rights and obligations attributed to it by the State in applicable legal provisions and regulations, particularly concerning enforced payment of levies, rates, service revenue and other credits.

Article 9: Powers of authority

1 – ERSAR exercises the necessary powers of authority for the carrying out of its duties, particularly through the carrying out of inspections, supervisory and auditing activities.

2 – The staff of ERSAR, when carrying out the activities foreseen in the previous paragraph, specifically enjoys the following prerogatives:

    a) Have free access to all facilities, infrastructure and equipment of the utilities;

    b) Obtain in any way, copies and extract of controlled documents, as well as collect samples, equipment and materials for the carrying out of analyses and tests, consultation, support and the attaching to reports, cases or minutes and, furthermore, carry out an examination of any relevant elements essential to the development of the aforementioned actions;

    c) Request, from any legal representative, worker or collaborator of the bodies subject to ERSAR regulation and also those who collaborate with the same bodies, clarification regarding facts or documents related to the purpose or objective of the inspection or audit and record their answers;

    d) Determine the suspension or termination of activities and the closure of facilities, following non-compliance with a precautionary measure requested by the managing board;

    e) Request the collaboration of competent bodies, specifically police and administrative authorities, when this is necessary for the carrying out of its functions.

3 – For the purposes of the previous paragraph, staff at ERSAR are accredited through the provision of an identification card approved and signed by the chairman of the managing board or, in his or her absence, or when unavailable to do so, through the joint signature of two members of the managing board, with external collaborators being accredited through a document issued for this purpose.

4 – Individuals mentioned in paragraph 2 should show the identification document referred to in the previous paragraph when carrying out their respective functions.

5 – It is incumbent on entities subject to intervention from ERSAR to arrange any conditions necessary to ensure the effectiveness of the activities carried out within the scope of its duties, particularly through the designation of interlocutors.

6 – Within the scope of the respective powers of supervision and whenever it is considered necessary due to the significant complexity or time spent on the analysis which the situation requires, ERSAR may contract experts and technicians to support and assist ERSAR staff, and providing to the same, under the scope of this service provision, the rights of access to relevant information. They are subject to the duty of confidentiality and restricted processing of information, under the terms applicable to ERSAR, following presentation of accreditation for this purpose.

Article 10: Sanctionary powers

ERSAR has the competence to deal with offences and apply the corresponding penalties and also any other applicable penalties to offences arising from legislation or regulations the implementation or supervision of which falls within its remit, as well as the results of non-compliance with its own resolutions, under the terms envisaged in law.

Article 11: Regulatory power

ERSAR has the competence to draw up and approve regulations with external effectiveness within the framework of its respective duties, without prejudice to others which may in the future be specified by law, specifically those concerning:

a) *Tariffs, under the terms laid down in Article 13;*
b) *Quality of service, particularly through defining minimal levels of quality and the compensation owed in the event of non-compliance;*
c) *Commercial relations, from the definition of relationship rules between utilities bulk and retail and between these latter and the corresponding users, specifically concerning conditions of access to and the contracting of the service, measuring, invoicing, paying and charging and provision of information and resolution of disputes, regulating the respective legal systems and the protection of uses of essential public services;*
d) *Regulatory procedures inherent to their relationship with the entities subject to their intervention, within the scope of its respective duties, thus embodying the manner and the period for carrying out the duties of the management body with regard to regulation;*
e) *Procedures for the approval of products in contact with drinking water, under the terms envisaged in Article 21 of Decree-Law No. 306/2007, of 27 August, altered by Decree-Law No. 92/2010, of 26 July.*

Article 12: Regulatory procedure

1 – Without prejudice to the consultation of ERSAR's advisory bodies, the approval or alteration of any regulation which contains provisions with external effectiveness, the approval of which is within its competence, is preceded by the holding of a public consultation period pursuant to law, for a period of not less than 30 working days, except for urgent situations which have duly been justified and which require the establishing of a shorter period. In that public consultation interested parties may present their comments and provide suggestions.

2 – For the purposes of the previous paragraph, ERSAR shall inform the members of the government responsible for the area of the environment and consumer protection, the holders of services, the utilities covered by the scope of the regulation and general consumer associations and the general public, of the draft regulation being drawn up, and provide access to the respective text and make it available on a page on its website.

3 – Once the period of public consultation has passed, ERSAR shall draw up and publish on its website a report analysing the comments and suggestions which have been made, where it will justify the decisions taken. The detailed justifications may be put in a supplementary document.

4 – ERSAR regulations with external effectiveness are published in the 2nd Series of the Portuguese Official Gazette and made available on ERSAR's website.

Article 13: Tariff regulations

1 – ERSAR shall approve tariff regulations for the water and waste services which will establish the following:

    a) Rules for the definition, fixing, revising and updating of tariffs for public drinking water supply, waste water management and solid waste management, in compliance with the following principles:

        i) Economic and financial recovery of service costs within an efficiency scenario;

        ii) Preservation of natural resources and the promotion of efficient behaviour by consumers;

        iii) Promotion of economic accessibility for final domestic users, particularly through social tariffs;

        iv) Promotion of equality in tariff structures, taking into consideration the size of the family unit, with special care given, in the case of domestic users, to large families, and giving special weighting to more just and efficient water abstractions, for all users;

        v) Stability and predictability by the regulated entities;

    b) Analytical accounting rules from the restricted perspective of an accounting separation of regulated activities themselves and any other activities that may be carried out by the utilities;

    c) Tariff convergence rules, of an exceptional nature, which allow for the transitory derogation of the principle of covering the costs which have been incurred in an efficiency scenario associated with service provision;

    d) Rules for the recovery of any excesses or insufficiencies in terms of charges incurred;

    e) Rules for the reporting of information to ensure compliance with applicable norms;

    f) Rules and procedures for inspection.

2 – The tariff regulations referred to in the previous paragraph shall be suitable for the specific situations of delegated management of services involving State ownership, under the terms of the law, which result in transfers to multimunicipal systems.

Article 14: Resolution of conflicts

1 – With regard to the exercising of its competences concerning the resolution of conflicts between regulated entities or between these and users, ERSAR should:

    a) Be aware of all complaints from users which come under its supervision and those which concern it, providing an answer and adopting any necessary measures with regard to these, thus recognising or not any alleged and invoked rights;

    b) Carrying out conciliation or promoting the use of arbitration in cases which from a procedural point of view are straightforward, streamlined and which tend to be of low significance, whenever this is envisaged in law or following a request from the interested parties.

2 – ERSAR should ensure that the procedures adopted under the terms of the previous paragraph are decided within a maximum period of 90 days counting from the date of reception of the request. This period may be extended for an equal period of time when ERSAR needs supplementary information or also for a longer period following agreement with the complainant.

3 – ERSAR may inspect the records of complaints made by users to the regulated entities.

4 – *Following its consideration of the complaints made and in accordance with each case, ERSAR may order or recommend that the utilities subject to its regulation take any necessary steps for the just reparation of the rights of the users.*

5 – *For the purposes envisaged in subparagraph b) of paragraph No. 1 of the present article, ERSAR shall promote the creation of new institutionalised arbitration centres, and may do this in collaboration with other entities, or enter into protocols with institutionalised arbitration centres, requiring in this case the specification of the logistical, financial, technical and human support to be provided for this purpose and, in addition, promote entities involved in the regulated sectors joining the aforementioned arbitration centres.*

CHAPTER III: Organisational structure

SECTION I: Listing of its bodies

Article 15: Bodies

The following are ERSAR bodies:

a) *The management board;*
b) *The sole auditor;*
c) *The advisory board;*
d) *The tariff board.*

SECTION II: The management board

Article 16: Function

The management board is the collegiate body responsible for the definition and implementation of ERSAR activity, as well as management of the respective services, in compliance with the law and the present bylaw.

Article 17: Composition

The management board consists of a chairman and two members, one of which may be designated as vice-chairman.

Article 18: Nomination

1 – *The members of the management board are chosen from among people of recognised standing, independence and technical competence, attitude, professional experience and suitable training for the carrying out of the respective functions.*

2 – *The nominees for members of the management board are nominated in a resolution of the council of ministers, following proposal from the member of the government responsible for the area of the environment.*

3 – *The nominations are preceded by hearings from the relevant parliamentary committee, at the request of the government, accompanied by a justification of the respective choices and the opinion of the public administration selection and recruitment committee regarding the suitability of their profile for the functions to be carried out, including compliance with any rules concerning incompatibility and any applicable impediments.*

4 – The resolution of the council of ministers which designates the members of the management board, suitably justified, is published in the Portuguese official gazette, along with a note regarding the academic and professional curricula of those chosen.

5 – In the event of simultaneous appointment of two or more members of the board of administration, the period of office of the respective mandates may not coincide, and there must be a difference between them of at least six months, by limiting the length of one or more the mandates, if necessary.

6 – Members of the management board may not be nominated in the period between the resignation of the government or the calling of parliamentary elections and the parliamentary investiture of the recently-chosen government, except if the vacancy of the post in question and the urgency of the choice has been verified, in which case the aforementioned choice or proposed choice which has not yet been effected depends on the confirmation of the recently-chosen government.

Article 19: Incompatibilities and impediments

1 – The members of the management board carry out their functions exclusively and may not, in particular:

> *a) Be members of sovereign bodies, for the autonomous regions or local government, nor carry out any other public or professional functions, except for teaching or research duties, provided that they are not remunerated;*
>
> *b) Maintain, directly or indirectly, any bond or relation, remunerated or not, with companies, groups of companies or other bodies involved in the activity of ERSAR or hold any shares or interests in the same;*
>
> *c) Maintain, directly or indirectly, any bond or relation, remunerated or not, with other entities, the activity of which may conflict with their duties and competencies.*

2 – In situations involving the termination of functions and for a period of two years the holders of management and equivalent posts may not establish any contractual bond or relationship with companies or other bodies regulated under the ERSAR activity. They are entitled, within the aforementioned period, to compensation equivalent to 1/2 of their monthly salary and are, in the case of non-compliance, obliged to return all accrued net earnings up to the maximum of three years, applying the updated coefficient resulting from the corresponding annual average variation rates for the consumer price index ascertained by Instituto Nacional de Estatística, I. P. (Portuguese statistics agency).

3 – Without prejudice to the provisions of the Framework Law for Regulatory Authorities, the members of the management board are also subject to the incompatibilities and impediments regime established for holders of senior public posts, and its specific aspects provided for by regulatory authorities.

Article 20: Length of mandate

The members of the management board are nominated for a period of six years, and this cannot be renewed.

Article 21: Termination of the mandate

1 – The members of the management board cannot be relieved of office before terminating the period of their mandate, except for cases provided for in this article.

2 – The mandate of the members of the management board ceases once it has run its course and also because of:

    a) Resignation, through a written statement presented to the member of the government responsible for the area of the environment;

    b) Death or permanent physical or mental incapacity or for a duration that is envisaged to go beyond the date of the period for which they have been chosen;

    c) Overriding incompatibility;

    d) Conviction, by final judicial decision which has the force of res judicata, of an intentional crime which raises the question of their suitability for holding their post;

    e) Serving a prison sentence;

    f) Dissolution of the management board or removal of its members under the terms of paragraphs 3 and 4, except for members of the management board who have explicitly maintained their mandates within the administrative body of the entity which will succeed it;

    g) Winding-up of ERSAR.

3 – The dissolution of the management board or removal of any of its members may only occur following a resolution of the council of ministers based on a justified reason.

4 – For the purposes of the provisions of the previous paragraph, the existence of a justified reason is understood to be the case whenever a case of serious misconduct, or individual or collective liability has been verified, which has been ascertained in an inquiry carried out by an independent body of the government. It is preceded by an opinion from the advisory board and a hearing of the relevant parliamentary committee, specifically in the case of:

    a) Serious or repeated contempt for the legal norms and bylaw, as well as ERSAR regulations and guidelines;

    b) Non-compliance of the duty to carry out functions exclusively or due to serious or repeated violation of the requirement for confidentiality;

    c) Substantial non-justified non-compliance with ERSAR's plan of activities or budget.

5 – In situations involving termination of their mandate through the respective period having passed and resignation, the members of the management board shall continue to carry out their functions until their effective substitution.

6 – In the case of vacancy due to one of the reasons provided for the previous paragraphs, the vacancy shall be filled within a maximum period of 45 days after confirmation.

Article 22: Statute of the members

1 – The gross monthly salaries of the members of the management board are established by the salary committee, under the general terms laid out in the Framework Law for regulatory authorities for the management boards of regulatory authorities.

2 – The members of the management board are entitled to the general social security scheme, except when they are holders of a legal public employment relationship for an indefinite period, in which case they may opt for the scheme of their place of origin.

3 – Situations associated with their functions or posts in bodies or other structures related with regulatory authorities involving members of the management board does not grant the right to any additional payment or any additional benefits and perks.

4 – Use of credit cards and other payment instruments, vehicles, communications and the right to social benefits by members of the management board is subject to the scheme defined for public managers.

Article 23: Salaries committee:

1 – Working with ERSAR is a salaries committee, under the terms defined in the Framework Law for Regulatory Authorities.

2 – The salaries committee is made up of three members, designated as follows:
   a) One chosen by the member of the government responsible for the area of finances;
   b) One chosen by the member of the government responsible for the area of the environment;
   c) One chosen by ERSAR, who has ideally held a post in one of the ERSAR bodies, or, if not chosen, co-opted by the aforementioned members in the previous subparagraphs.

Article 24: Duties of the management board

1 – The duties of the management board as regards regulation and supervision are:
   a) Issuing opinions, studies, information and draft legislation at the request of the government and on their own initiative in matters related to the scope of their respective duties, for the clarification of rules of functioning of the water and waste services, and monitor the drawing up and application of the respective legislation;
   b) Approving regulations with external effectiveness as provided for in the law and which are necessary for carrying out the duties of ERSAR;
   c) Drawing up or evaluation on tariffs for State and municipal holdings under the terms laid out in the respective legal schemes;
   d) Issuing recommendations and codes of best practices for any matter falling within the remit of ERSAR within the scope of its respective duties;
   e) Taking any decisions necessary in carrying out the duties of ERSAR and issuing instructions on aspects related to the scope of those duties;
   f) Issuing opinions under the scope of the attribution and contracting of multimunicipal concessions, the setting up of intermunicipal systems, delegation of municipal services, public contracting procedures for the selection of private partners and the attribution of municipal concessions, the respective contracting, as well as those for concessions, the entering into of partnership contracts between the municipalities and the State and management contracts concerning this, and the alteration and cessation of contracts and also public service regulations. These opinions are published under the terms of subparagraph b) of paragraph 3 of Article 50 of the present statues and sent to all interested parties;
   g) Raise the matter of the reconsideration of contractual clauses with the holder of the services, which deal with public interest and when these are managed by contract;
   h) Specify the carrying out of inspection and auditing activities for sector systems, independently of their ownership, management model or services provided.
   i) Specify the carrying out of inspections on the supply systems and supervision of the laboratories for analysing drinking water, within the scope of the control of the quality of drinking water;
   j) Exercise sanctionary power, under the terms defined in the applicable legislation;
   k) Request any cautionary measures and those of a similar nature, or any other form of proceedings regarding matters which may jeopardise the balance of the sectors and ensure the defence of consumer rights and those which are shown to be necessary for the prevention or cessation of activities contrary to that provided for in legislation and compliance with that which it has the duty to supervise;

l)   Enter into cooperation or collaboration protocols and establish mechanisms of association with other entities of public or private law, either national or international, when this is shown to be necessary or important for the carrying out of the duties of ERSAR;

m)   Coordinate and carry out the collection and dissemination of important information regarding the regulatory model, the sectors for the public drinking water supply, waste water management and solid waste management services and the respective utilities;

n)   Promote research, innovation and the carrying out of studies on matters related to its duties;

o)   Draw up the annual regulation and supervision report;

p)   Carry out all and any other acts necessary for the performance of the duties of ERSAR for which there is no other competent body.

2 –   The duties of the internal management of the management board are:

a)   Manage ERSAR's activity and its services;

b)   Draw up the annual activity plans and ensure their respective implementation, monitoring and assessment;

c)   Draw up the draft budget, under the terms of applicable legislation;

d)   Propose any necessary budgetary alterations, without prejudice to the mechanisms for budgetary approval provided for in the framework law for regulatory authorities;

e)   Draw up the annual activities report and accounts;

f)   Draw up the authority's balance sheet, under the terms of the applicable legislation;

g)   Exercise its powers regarding staff administration, management and discipline, as well as carry out any other acts relating to this, under the terms provided for in law, in its bylaw and in the internal regulations to be approved;

h)   Approve the internal regulations necessary for the carrying out of its duties;

i)   Carry out any other acts of management resulting from the application of the present bylaw and other applicable legislation which is seen as necessary to the proper functioning of its services;

j)   Accompany and systematically assess the activity carried out, particularly that promoting a rational use of available resources in terms of maximising results;

k)   Nominate representatives from ERSAR for external bodies;

l)   Establish powers-of-attorney for ERSAR, in and out of court, which also have the power of appointing substitutes.

3 –   The management board also has competence to carry out day-to-day management acts necessary for the good functioning of ERSAR and carry out all the other competences granted to it in it bylaw and in the framework law for regulatory authorities either delegated or subdelegated in it.

4 –   All the entities subject to ERSAR's activity taking decisions with do not comply with the recommendations or opinions of ERSAR as envisaged in subparagraph d) and f) paragraph No. 1, have to explicitly justify their decision, with a detailed statement of the reasons based on fact and in law which justified the grounds for the action.

5 –   The decisions referred to in the previous paragraph were obligatorily subject to publication on ERSAR's website, that of the decision-making entity, as well as in a suitable official publication at a national, regional or local level, within a period of 15 days.

Article 25: Duties of the chairman of the management board

1 –   The chairman of the management board is required to:

a)   Convene and chair meetings of the management board, guide its operations and promote compliance with its respective decisions;

b)   *Co-ordinate the activity of the management board and its relations with other ERSAR bodies and services;*

c)   *Co-ordinate its relations with the government, with other public bodies and with holding and management bodies;*

d)   *Request the convening of the advisory board to consider appropriate matters;*

e)   *Exercise the competences which have been delegated to it by the management board;*

f)   *Exercise other competences envisaged in this bylaw or in law.*

2 –   *The chairman of the management board is substituted, when absent or unable to attend, by the vice-chairman, when this post exists, or by the member indicated by him or her and, in their absence, by the longest standing member.*

Article 26: Delegation of duties

1 –   *With the exception of the competences provided for in subparagraphs b), c) and j) of paragraph No. 1 of Article 24, the management board and its chairman may delegate its respective duties, following its decision or order, depending on the case, in one or more of its members and authorise the subdelegation of these powers be made to administrators or employees of ERSAR, in each case establishing the respective limits, conditions and control mechanisms.*

2 –   *The provisions of the previous paragraph do not prejudice the duty which is incumbent upon all the members of the management board to be aware of and generally monitor ERSAR affairs and make rulings on these, nor the power of the board of management to call back the delegated, subdelegated and mandated powers or to revoke the acts carried out by the delegated, subdelegated and mandated party under the scope of its delegation, subdelegation or mandate, whenever it considers this appropriate for the carrying out of the duties of ERSAR.*

Article 27: Operation

1 –   *The management board normally meets once a week and exceptionally whenever the chairman convenes it, either through his initiative or following a request from the members.*

2 –   *Abstaining is not an option when voting, but explanations of votes may be made.*

3 –   *The minutes of each meeting must be approved and signed by all the members present, and members disagreeing with the contents of the decisions taken may have their respective non-approval recorded in the minutes.*

Article 28: Representation, substitution and powers to bind

1 –   *ERSAR is represented, particularly in court or in the practice of legal acts, by the chairman of the management board, by two of its members or by an attorney especially constituted for such purpose.*

2 –   *ERSAR binds itself with the joint signature of two of the members of the management board, with one of them being the chairman or his or her substitute.*

3 –   *Without prejudice to the provisions of the previous paragraph, as far as day-to-day management is concerned, as defined following the decision of the management board, ERSAR may bind itself through the signature of one member of the management board or any employee of ERSAR in carrying out powers subdelegated in him/her.*

4 –   *Without prejudice to the provisions of the previous paragraphs, ERSAR may also bind itself through the signature of its attorneys, in the restricted scope of the powers which have been granted within the respective mandate.*

SECTION III: Sole Auditor

Article 29: Function

The sole auditor is responsible for monitoring the legality, accuracy and sound financial and asset management of ERSAR and by the exercise of consultative competences in this field.

Article 30: Nomination

1 – The sole auditor is nominated by order of the members of the government responsible for the areas of finance and the environment.

2 – The sole auditor must be a statutory auditor or a firm of statutory auditors.

Article 31: Incompatibilities and impediments

The sole auditor chosen may not maintain any binding labour relationship with the State nor maintain, directly or indirectly, any contractual bond or relation, remunerated or not, with companies, groups of companies or other bodies which are the targets of the activity of ERSAR or hold any shares or interests in the same, nor other entities whose activity may conflict with its duties and competencies;

Article 32: Length of mandate

The sole auditor is nominated for a period of four years, and this mandate cannot be renewed.

Article 33: Statute of the sole auditor

1 – The sole auditor should be independent in carrying out its functions, and not be subject to instructions or guidelines, and be governed by the legal provisions respecting the performance of the activity of statutory auditor.

2 – The sole auditor has the right to a monthly salary, paid 12 times a year, to the value of 1/4 of the monthly salary established for the chairman of the management board.

Article 34: Duties of the sole auditor

1 – The sole auditor should regularly monitor and control compliance with applicable laws and regulations, budgetary execution, the economic, financial, assets and accounting situation of ERSAR and carry out other competences attributed to it under the terms of the law, specifically the consultative competences envisaged in the framework law for regulatory authorities.

2 – The sole auditor should check the quality of the systems regarding performance indicators for efficiency, effectiveness and quality, which reflects the set of activities carried out and the results obtained, as well as annually assess the results obtained by ERSAR in terms of the available resources, the conclusions of which are reported to the member of the government responsible for the area of the environment, in accordance with the provisions of Article 39 of the framework law for regulatory authorities.

SECTION IV: Advisory Board

Article 35: Function, duties and composition

1 – The advisory board is the consultation body for the general areas of operation of ERSAR, thus ensuring the participation of representatives of the main interested parties involved in the activities of the regulated sectors involving water and waste services.

2 – The advisory board should contribute to the formulation of public policies for the sectors and issue an opinion on:

    a)   The annual plan and activities report and accounts;

    b)   The regulatory model;

    c)   Other issues the consideration of which are submitted to it by the management board.

3 – The advisory board should also present, through its own initiative, suggestions and proposals to the management board aimed at promoting the improvement of the sectors and the activities of ERSAR within the framework of its respective duties.

4 – The ERSAR advisory board is chaired by a personality of recognised merit, nominated by the member of the government responsible for the area of the environment.

5 – The advisory board also includes the following members:

    a)   The director-general for local authorities;

    b)   The director-general for economic activities;

    c)   The director-general for the consumer;

    d)   The director-general for health;

    e)   The chairman of the Portuguese environment agency;

    f)   A representative from the regional co-ordination and development committees at the level of the chairman or vice chairman, in a rotating system;

    g)   A representative from each from the autonomous regions;

    h)   A representative from the national association of Portuguese municipalities;

    i)   Four representatives from drinking water supply and waste water managements utilities at a municipal ownership level, through direct, delegated, partnership or concession management, two of which should represent public entities and two the private entities;

    j)   Three representatives from solid waste management utilities at a municipal ownership level, through direct, delegated, partnership or concession management, one of which should represent public entities and two the private entities;

    k)   One representative from drinking water supply and solid waste water utilities at a State ownership level, through direct, delegated or concession management;

    l)   One representative from solid waste management utilities at a State ownership level, through direct, delegated or concession management;

    m) One representative from utilities of specific waste streams;

    n)   Two representatives from consumer associations operating at a national level;

    o)   Four representatives from associations representing economic activities which operate at a national level;

    p)   Four representatives from significant technical and professional associations which operate in the sectors;

    q)   Two representatives from non-governmental environmental organisations which operate at a national level.

6 – Also forming part of the advisory board are specialists from the public drinking water supply, waste water management and solid waste management sectors, with a maximum number of three members, nominated by order of the member of the government responsible for the area of the environment, following a proposal to this effect by the chairman of the advisory board.

7 – The members which are referred to in sub paragraphs a) to e) of paragraph No. 5 carry out this mandate due to their respective functions.

8 – The carrying out of the post of chairman of the advisory board and the specialists mentioned in paragraph No. 6 remunerated through attendance fees, of value to be defined through internal regulation, which may not exceed the limit of two payments corresponding to the allowance paid by ERSAR for help with travels costs within national territory.

9 – *The members of the advisory board who are representatives of non-profit non-governmental organisations may request compensation for travel and subsistence costs, through attendance fees, which cannot be accumulated with the allowances mentioned in the previous paragraph, at a value equivalent to the allowance given by ERSAR to help with travel costs in national territory, under the terms to be defined in an internal ERSAR regulation.*

10 – *The advisory board may set up specialised sections relating to water and waste services or in specific areas, under the terms to be defined in the respective internal regulation.*

11 – *The represented entities included in each of the categories referred to in sub paragraphs i) to q) paragraph No. 5, may agree amongst themselves to share their representative mandate, selecting two or more representatives, to be defined through internal regulation, who may substitute them halfway through their mandate.*

12 – *Should there be no confederated structures at the national level, to link the bodies which may be represented, and there are difficulties in establishing a platform of understanding concerning their representation, the following procedure shall be adopted:*

   a) *The chairman of the advisory board, based on impartial criteria concerning representation, shall draw up a proposal which will indicate one or more bodies to join the advisory council mentioned in each sub paragraph of paragraph No. 5;*

   b) *The proposal mentioned in the previous paragraph shall be submitted to all those bodies that could be represented so that within a period of 30 working days they can issue a statement on this. They may submit an alternative proposal, on the condition that, if they do not do so, it is considered that they have accepted the original proposal;*

   c) *In the event of acceptance by a simple majority of the bodies consulted, the chairman of the advisory council shall then issue a formal invitation to the body in question to specify its representatives;*

   d) *Should the proposal be refused by simple majority of the bodies consulted, the chairman of the advisory board shall, in a reasoned manner, and bearing in mind the alternative proposals submitted, decide which body or bodies shall be named as representatives to form part of the advisory body, with it being possible for this mandate to be carried out on a rotating basis.*

13 – *The advisory board shall normally meet at least twice a year and be convened by its chairman.*

14 – *Exceptionally, the advisory board may meet following its convening by its chairman, on his or her initiative, at the request of at least one-third of its members, or at the request of the chairman of the management board.*

15 – *The members of the management board may participate, without the right to vote, in the meetings of the advisory board.*

16 – *The advisory board shall approve its internal regulations.*

Article 36: Length of mandate

1 – *Members of the advisory board are nominated for a period of three years, without prejudice to the fact that they may be substituted at any moment by the bodies which have nominated them.*

2 – *The effective members, referred to in sub paragraphs f) to q) of paragraph No. 5 of the previous article, may be substituted by supplementary members, to be specified in the appointment document of the effective member.*

SECTION V: Tariff board

Article 37: Function, duties and composition

1 – The tariff board is the specific advisory board for ERSAR functions relating to tariffs and prices.

2 – It is incumbent upon the tariff board to:
  a) Issue an opinion on the proposal for tariff regulation and its revisions;
  b) Annually issue an opinion regarding the balance sheet of the economic regulation cycle;

3 – The tariff board is chaired by the chairman of the advisory board and has the following composition:
  a) One representative from the directorate general for local authorities;
  b) One representative from the directorate general for economic activities;
  c) One representative from the directorate general for the consumer;
  d) One representative from the Portuguese environment agency;
  e) One representative from the national association of Portuguese municipalities;
  f) Four representatives from drinking water supply and waste water managements utilities at a municipal ownership level, through direct, delegated, partnership or concession management, two of which should represent the public entries and two the private entities;
  g) Three representatives from solid waste management utilities at a municipal ownership level, through direct, delegated, partnership or concession management, one of which should represent the public entries and two the private entities;
  h) One representative from drinking water supply and waste water management utilities at a State ownership level, through direct, delegated or concession management;
  l) One representative from solid waste management utilities at a State ownership level, through direct, delegated or concession management;
  j) One representative from utilities of specific waste streams;
  k) Two representatives from consumer associations operating at a national level;

4 – The holding of posts on the tariff board is not remunerated.

5 – The tariff board shall normally meet once a year and be convened by its chairman.

6 – Exceptionally, the tariff board may meet following its convening by its chairman, on his or her initiative, at the request of at least one-third of its members, or at the request of the chairman of the management board.

7 – The members of the management board may participate, without the right to vote, in the meetings of the tariff board.

8 – The tariff board shall approve its internal regulations.

Article 38: Length of mandate
The length of the mandate of the members of the tariff board is according to the rules stated in Article 36.

CHAPTER IV: Services and staff

Article 39: Operational and support services

1 – ERSAR makes use of operational and technical and administrative support services, which are indispensable in the performance of its duties.

2 – The internal services regulation, which specifies its internal organisation, careers, administrative posts in ERSAR and pay level is approved by the management board.

Article 40: Personnel regime

1 – ERSAR personnel are governed by the legal scheme relating to an individual contract of employment, with the provisos provided for in this bylaw.

2 – ERSAR may be a party to a collective labour agreement.

3 – Conditions for the recruitment and selection of employees, their work performance and discipline are defined in an internal regulation approved by the management board, observing the following general principles:

a)   Employment vacancies are published on ERSAR's website;

b)   Equality of conditions and opportunities for candidates;

c)   Application of methods and rigorous objective criteria for assessment and selection;

d)   Justification of decisions taken.

4 – The adoption of the legal scheme relating to an individual contract of employment does not do away with the requirements and limitations resulting from serving the public interest, specifically regarding the accumulations and incompatibilities legally established for workers in the public sectors and those provided for in the framework law for regulatory authorities.

5 – Performance assessment of ERSAR workers is carried out through applying criteria and guidelines established in relation to:

a)   Principles and objectives, as well as the existence of evaluation systems for workers, administrators and organisational units, functioning in an integrated manner;

b)   Performance assessment based on the confrontation between established objectives and results obtained and, in the case of the administrators and employees, in addition to the duties shown and being developed;

c)   Differentiation of performances through the establishing of a minimum number of assessment mentions and maximum percentages for the awarding of the highest mentions.

6 – ERSAR's performance assessment system, which follows the provisions of the previous paragraph, is defined in an internal regulation approved by the management board.

7 – The employees provided for in paragraph No. 1 are registered in the general social security scheme for employees, except for the right to opt for continuing to be registered in the Caixa Geral de Aposentações (Civil servants pension fund) for workers with a legal relationship of public employment.

Article 41: Other personnel

The employees which carry out public functions, as well as any employees, staff or public or private company administrators, may carry out functions in ERSAR or in any of its bodies through recourse to the legally applicable means concerning mobility.

Article 42: Contracting of external services and cooperation protocols

ERSAR may contract, under outsourcing format, the cooperation of companies or specialists in order to draw up studies, opinions, monitor audits and inspection activities or other tasks necessary for the carrying out of its functions.

Article 43: Duties of confidentiality, care and circumspection

1 – Members of ERSAR bodies, as well as staff and service providers and their collaborators, are subject to the duties of care and confidentiality concerning all matters entrusted to them or that they have knowledge of arising from the carrying out of their functions.

2 – *The members of the management board of ERSAR may not make statements or comment on proceedings in progress or specific issues relating to regulated bodies, except in defence of their honour or in order to carry out another legitimate interest.*

3 – *The requirement for circumspection does not apply to proceedings which have been concluded, as well as the provision of information which seeks to implement rights or legitimate interests, particularly that of access to information.*

CHAPTER V: Assets, budgetary and financial regime

Article 44: Assets

1 – *ERSAR's patrimony is made up of assets, rights and obligations of an economic nature, provided for by the State or acquired by it.*

2 – *ERSAR is governed by the regimes concerning public immovable property, movable property and the State pool of vehicles, with regard to the goods assigned it by the same, and by private law in relation to its other assets.*

Article 45: Applicable regime

1 – *ERSAR has, as regards its financial and asset management, its own autonomy as provided for in this bylaw and in the framework law for regulatory authorities.*

2 – *The rules for public accounting, the autonomous regime for funds and services, particularly the norms concerning the authorisation of expenses, the transfer and use of the net results for the year and the blocking of funds regarding the part which does not depend on allocations from the State budget are not applicable to ERSAR.*

Article 46: Revenue

1 – *The following constitute own income for ERSAR:*
  a) *The rates and contributions charged to utilities of drinking water supply, waste water management and solid waste management services concerning structural, economic and the quality of service regulation activity.*
  b) *The rates and contributions charged to utilities of water supply as the national authority for the regulation of the quality of drinking water;*
  c) *The rates due for procedures involving approval, authorisation or recognition for which ERSAR is responsible;*
  d) *The income from penalties applied due to infringements that it is ERSAR's duty to sanction;*
  e) *Income arising from the services provided by ERSAR;*
  f) *Income arising from the profit from, sale or leasing of its own assets, or that resulting from cash-related financial investments;*
  g) *Subsidies, funding, grants and donations provided by any national or foreign bodies;*
  h) *Any income which by law, contract or other form should be attributed to it.*

2 – *The requirements, criteria for levying and the value of the rates and contributions provided for in subparagraphs a) to c) of the previous paragraph are defined by ministerial order of members of the government responsible for the areas of finance and the environment.*

Article 47: Expenses

ERSAR expenses are those items that result from charges arising from the carrying out of its respective duties and, as well, the contributions legally attributed to it within the scope of the regime for the funding of the competition authority.

Article 48: Accounting, accounts and treasury

1 – ERSAR shall use the accounting standards system.

2 – The rendering of accounts is essentially governed by the law on organisation and procedure of the court of auditors and the respective regulatory provisions.

3 – The State treasury scheme is applicable to ERSAR and, in particular, the principle and the rules concerning the treasury unit.

4 – ERSAR shall draw up and each year update the respective inventory of its immovable property, under the terms of the legal scheme concerning public immovable property.

*5 – The annual net profits of ERSAR shall be retained for the following year and should be utilised to establish or reinforce the reserves aimed at developing specific actions for the benefit of the sectors, particularly actions involving technical capacity building for the utilities and other stakeholders **of the sectors**.*

CHAPTER VI: Independence, liability and legal control

Article 49: Independence

1 – ERSAR is independent in the carrying out of its functions and is not subject to supervision or governmental guardianship, under the terms of this bylaw and the framework law for regulatory authorities, without prejudice to the provisions of the following paragraphs.

2 – The members of the government may not administer recommendations or issue directives to the administrative bodies of ERSAR regarding its regulatory activity or concerning the priorities it should adopt in its respective operation.

3 – The member of the government responsible for the area of the environment may request information from its bodies regarding the carrying out of its annual and multi-annual activities plan, as well as its budgets and the respective multi-annual plans.

4 – Prior approval is required on the part of the members of the government responsible for the area of finances and for the area of the environment, within a period of 60 days after its reception, for the budget regarding the following financial year, the corresponding multi-annual plan, as well as the management report and the balance sheet and the accounts for the previous financial year.

5 – The approvals provided for in the previous paragraph may only be refused following a decision based on their illegality or their prejudice of the objectives of ERSAR or in the public interest, or also as an unfavourable opinion issued by the advisory board.

6 – Once the period envisaged for paragraph No. 4 has expired without an express decision having been issued, the respective documents are considered to have been tacitly approved.

7 – Prior authorisation is also required on the part of the members of the government responsible for the areas of finance and the environment, otherwise they will be rendered null and void, for the following:

a)   The acceptance of donations, inheritances or bequests;

b)   The acquisition or disposal of immovable property, under the terms of the law.

Article 50: Provision of information

1 – ERSAR shall draw up and send to parliament and the government a detailed report on its respective functioning and regulation and supervision activity, and that report will be published on its website.

2 – Whenever requested to do so, the members of the management board of ERSAR shall present themselves to the competent parliamentary committee, in order to provide information and clarification concerning the respective activity.

3 – ERSAR has made available on its website all important data for the sectors and concerning its activity, namely:

a)   The composition of its statutory bodies, including biographical, curricular and earning status of the respective members;

b)   The legislation and regulations relevant to the regulated sectors, the framework law for regulatory authorities, its regulatory instruments, this bylaw, opinion issued under the terms of subparagraph No. 1 of Article 24 of the present bylaw and its internal regulations;

c)   The annual reports of the water and waste services in Portugal;

d)   Management instruments, namely:

i)   Activities plans and budgets;

ii)   Activities reports and approved accounts, including the respective balance sheets.

Article 51: Liability

1 – ERSAR, the members of its bodies and its employees are held liable in civil, criminal, disciplinary and financial proceedings for any acts and emissions which they carry out in the exercise of their duties, under the terms of the applicable legislation.

2 – The financial responsibility is undertaken by the court of auditors, under the terms of the respective legislation.

3 – If they are made defendants by third parties, under the terms of paragraph No. 1, the members of the ERSAR bodies and its employees have the right to legal support assured by the regulatory authority, without prejudice to rights of recourse in general terms.

Article 52: Judicial review

1 – Questions concerning the appeal, revision and the carrying out of decisions, orders or other measures legally susceptible to dispute which are taken by ERSAR, in administrative infraction proceedings, are placed in the competition, regulation and supervision court, under the terms of the applicable legislation, with all the other acts of an administrative nature carried out by the bodies of ERSAR subject to administrative jurisdiction, in accordance with the respective legislation.

2 – ERSAR has the legitimate right to appeal decisions taken in judicial review proceedings which permit appeal.

Index

IWA Publishing's authorised EU representative for General Product
Safety Regulations is Diane D'Arras, 15 rue Duret, 75116 Paris,
France, e-mail: safety@iwap.co.uk.

Printed and bound by CPI Group (UK) Ltd, Croydon, CR0 4YY
27/03/2026
02079977-0001